Firefighter Fatalities in the United States in 2010

Prepared by

U.S. Department of Homeland Security

Federal Emergency Management Agency

U.S. Fire Administration

National Fire Data Center

and

The National Fallen Firefighters Foundation

www.firehero.org

In memory of all firefighters
who answered their last call in 2010
To their families and friends
To their service and sacrifice

"We will never forget"

Cover photo courtesy of Rae Stewart.

U.S. Fire Administration
Mission Statement

We provide National leadership to foster a solid foundation for local fire and emergency services for prevention, preparedness and response.

TABLE OF CONTENTS

ACKNOWLEDGMENTS

This study of firefighter fatalities would not have been possible without the cooperation and assistance of many members of the fire service across the United States. Members of individual fire departments, chief fire officers, wildland fire service organizations such as the U.S. Forest Service (USFS), the National Park Service (NPS), the Bureau of Land Management (BLM), the Bureau of Indian Affairs (BIA), the U.S. Fish and Wildlife Service (FWS), as well as the U.S. Department of Justice (DOJ), the National Fire Protection Association (NFPA), and many others contributed important information to this report.

The National Fallen Firefighters Foundation (NFFF) was responsible for compilation of a large portion of the data used in this report and the incident narrative summaries found in Appendix A.

The ultimate objective of this effort is to reduce the number of firefighter deaths through an increased awareness and understanding of their causes and how they can be prevented. Firefighting, rescue, and other types of emergency operations are essential activities in an inherently dangerous profession, and unfortunate tragedies do occur. These are the risks all firefighters accept every time they respond to an emergency incident. However, the risks can be greatly reduced through efforts to improve training, emergency scene operations, and firefighter health and safety initiatives.

BACKGROUND

For 34 years, the U.S. Fire Administration (USFA) has tracked the number of firefighter fatalities and conducted an annual analysis. Through the collection of information on the causes of firefighter deaths, the USFA is able to focus on specific problems and direct efforts toward finding solutions to reduce the number of firefighter fatalities in the future. This information is also used to measure the effectiveness of current programs directed toward firefighter health and safety.

Several programs have been funded by USFA in response to this annual report. For example, USFA has sponsored significant work in the areas of general emergency vehicle operations safety, fire department tanker/tender operations safety, firefighter incident scene rehabilitation, and roadside incident safety. The data developed for this report are also widely used in other firefighter fatality prevention efforts.

One of USFA's main program goals is a 25-percent reduction in firefighter fatalities in 5 years and a 50-percent reduction within 10 years. The emphasis placed on these goals by USFA is underscored by the fact that these goals represent one of the five major objectives that guide the actions of USFA.

In addition to the analysis, USFA, working in partnership with the NFFF, develops a list of all onduty firefighter fatalities and associated documentation each year. If certain criteria are met, the fallen firefighter's next of kin, as well as members of the individual's fire department, are invited to the annual National Fallen Firefighters Memorial Weekend Service. The service is held at the National Emergency Training Center (NETC) in Emmitsburg, MD, during Fire Prevention Week in October of each year. Additional information regarding the Memorial Service can be found at www.firehero.org or by calling the NFFF at (301) 447-1365.

Other resources and information regarding firefighter fatalities, including current fatality notices, the National Fallen Firefighters Memorial database, and links to the Public Safety Officers' Benefit (PSOB) Program can be found at www.usfa.dhs.gov/fireservice/fatalities/

INTRODUCTION

This report continues a series of annual studies by the U.S. Fire Administration (USFA) of onduty firefighter fatalities in the United States.

The specific objective of this study is to identify all onduty firefighter fatalities that occurred in the United States and its protectorates in 2010 and to analyze the circumstances surrounding each occurrence. The study is intended to help identify approaches that could reduce the number of firefighter deaths in future years.

Who is a Firefighter?

For the purpose of this study, the term firefighter covers all members of organized fire departments with assigned fire suppression duties in all 50 States, the District of Columbia, and the Territories of Puerto Rico, the Virgin Islands, American Samoa, the Commonwealth of the Northern Mariana Islands, and Guam. It includes career and volunteer firefighters; full-time public safety officers acting as firefighters; fire police; State, territory, and Federal government fire service personnel, including wildland firefighters; and privately employed firefighters, including employees of contract fire departments and trained members of industrial fire brigades, whether full or part time. It also includes contract personnel working as firefighters or assigned to work in direct support of fire service organizations (i.e., air-tanker crews).

Under this definition, the study includes not only local and municipal firefighters, but also seasonal and full-time employees of the U.S. Forest Service (USFS), the National Park Service (NPS), the Bureau of Land Management (BLM), the Bureau of Indian Affairs (BIA), the U.S. Fish and Wildlife Service (FWS), and State wildland agencies. The definition also includes prison inmates serving on firefighting crews; firefighters employed by other governmental agencies, such as the U.S. Department of Energy (DOE); military personnel performing assigned fire suppression activities; and civilian firefighters working at military installations.

What Constitutes an Onduty Fatality?

Onduty fatalities include any injury or illness sustained while on duty that proves fatal. The term "on duty" refers to being involved in operations at the scene of an emergency, whether it is a fire or nonfire incident; responding to or returning from an incident; performing other officially assigned duties such as training, maintenance, public education, inspection, investigations, court testimony, and fundraising; and being on call, under orders, or on standby duty except at the individual's home or place of business. An individual who experiences a heart attack or other fatal injury at home while he or she prepares to respond to an emergency is considered on duty when the response begins. A firefighter that becomes ill while performing fire department duties and suffers a heart attack shortly after arriving home or at another location may be considered on duty since the inception of the heart attack occurred while the firefighter was on duty.

On December 15, 2003, the President of the United States signed into law the Hometown Heroes Survivors Benefit Act of 2003. After being signed by the President, the Act became Public Law 108-182. The law presumes that a heart attack or stroke are in the line of duty if the firefighter was engaged in nonroutine stressful or strenuous physical activity while on duty and the firefighter becomes ill while on duty or within 24 hours after engaging in such activity. The full text of the law is available at http://frwebgate.access.gpo.gov/cgi-bin/getdoc.cgi?dbname=108_cong_public_laws&docid=f:publ182.108.pdf

The inclusion criteria for this study have been affected by this change in the law. Previous to December 15, 2003, firefighters who became ill as the result of a heart attack or stroke after going off duty needed to register a complaint of not feeling well while still on duty in order to be included in this study. For firefighter fatalities after December 15, 2003, firefighters will be included in this report if they became ill as the result of a heart attack or stroke within 24 hours of a training activity or emergency response. Firefighters who became ill after going off duty where the activities while on duty were limited to tasks that did not involve physical or mental stress will not be included.

A fatality may be caused directly by an accidental or intentional injury in either emergency or nonemergency circumstances, or it may be attributed to an occupationally related fatal illness. A common example of a fatal illness incurred on duty is a heart attack. Fatalities attributed to occupational illnesses also include a communicable disease contracted while on duty that proved fatal when the disease could be attributed to a documented occupational exposure.

Firefighter fatalities are included in this report even when death is considerably delayed after the original incident. When the incident and the death occur in different years, the analysis counts the fatality as having occurred in the year in which the incident took place.

There is no established mechanism for identifying fatalities that result from illnesses such as cancer that develop over long periods of time and which may be related to occupational exposure to hazardous materials or toxic products of combustion. It has proved to be very difficult to provide a complete evaluation of an occupational illness as a causal factor in firefighter deaths due to the following limitations: the exposure of firefighters to toxic hazards is not sufficiently tracked; the often delayed long-term effects of such toxic hazard exposures; and the exposures firefighters may receive while off duty.

Sources of Initial Notification

As an integral part of its ongoing program to collect and analyze fire data, the USFA solicits information on firefighter fatalities directly from the fire service and from a wide range of other sources. These sources include the Public Safety Officers' Benefit (PSOB) Program administered by the U.S. Department of Justice (DOJ), the National Institute for Occupational Safety and Health (NIOSH), the Occupational Safety and Health Administration (OSHA), the Department of Defense (DOD), the National Interagency Fire Center (NIFC), and other Federal agencies.

The USFA receives notification of some deaths directly from fire departments, as well as from such fire service organizations as the International Association of Fire Chiefs (IAFC), the International Association of Fire Fighters (IAFF), the National Fire Protection Association (NFPA), the National Volunteer Fire Council (NVFC),

State fire marshals, State fire training organizations, other State and local organizations, fire service Internet sites, news services, and fire service publications.

Procedure for Including a Fatality in the Study

In most cases, after notification of a fatal incident, initial telephone contact is made with local authorities by the USFA to verify the incident, its location, jurisdiction, and the fire department or agency involved. Further information about the deceased firefighter and the incident may be obtained from the chief of the fire department, designee over the phone, or by other forms of data collection. After basic information is collected, a notice of the firefighter fatality is posted at the National Fallen Firefighters Memorial site in Emmitsburg, MD, the USFA website, and a notice of the fatality is transmitted by electronic mail to a large list of fire service organizations and fire service members.

Information that is routinely requested from fire departments that have experienced a fatality includes National Fire Incident Reporting System (NFIRS)-1 (incident) and NFIRS-3 (fire service casualty) reports; the fire department's own incident and internal investigation reports; copies of death certificates and autopsy results; special investigative reports; law enforcement reports; photographs and diagrams; and newspaper or media accounts of the incident. Information on the incident may also be gathered from NFPA or NIOSH reports.

After obtaining this information, a determination is made as to whether the death qualifies as an onduty firefighter fatality according to the previously described criteria. With the exception of firefighter deaths after December 15, 2003, the same criteria were used for this study as in previous annual studies. Additional information may be requested by USFA, either through follow-up with the fire department directly, from State vital records offices, or other agencies. The final determination as to whether a fatality qualifies as an onduty death for inclusion in this statistical analysis is made by the USFA and the Foundation's criteria as a line-of-duty death (LODD) for inclusion in the annual National Fallen Firefighters Memorial Service is made by the National Fallen Firefighters Foundation (NFFF).

2010 FINDINGS

Eighty-seven firefighters died while on duty in 2010, continuing a second year of substantially fewer firefighter deaths in the United States, where, during the previous 6-year period of 2004–2009, the average number of annual onduty firefighter deaths was 112.

The 2010 total includes 15 firefighters who died under circumstances as a result of inclusion criteria changes resulting from the Hometown Heroes Act of 2003. When not including these fatalities in a trend analysis, the 2010 total of 72 firefighter fatalities was the lowest number of firefighter losses on record over the past 34 years.

An analysis of multiyear firefighter fatality trends needs to acknowledge the changes from the Home-

town Heroes Survivors Benefit Act of 2003; therefore, some graphs and charts either will or will not indicate the Hometown Heroes portion of the total. However, this does not diminish the sacrifices made by any firefighter who dies while on duty or the sacrifices made by his/her family and peers.

Moreover, when conducting multiyear comparisons of firefighter fatalities in this report, the losses that were the result of the attacks on the World Trade Center (WTC) in New York City on September 11, 2001, are sometimes also set apart for illustrative purposes. This action is by no means a minimization of the supreme sacrifice made by these firefighters.

Figure 1. Onduty Firefighter Fatalities (1977–2010).

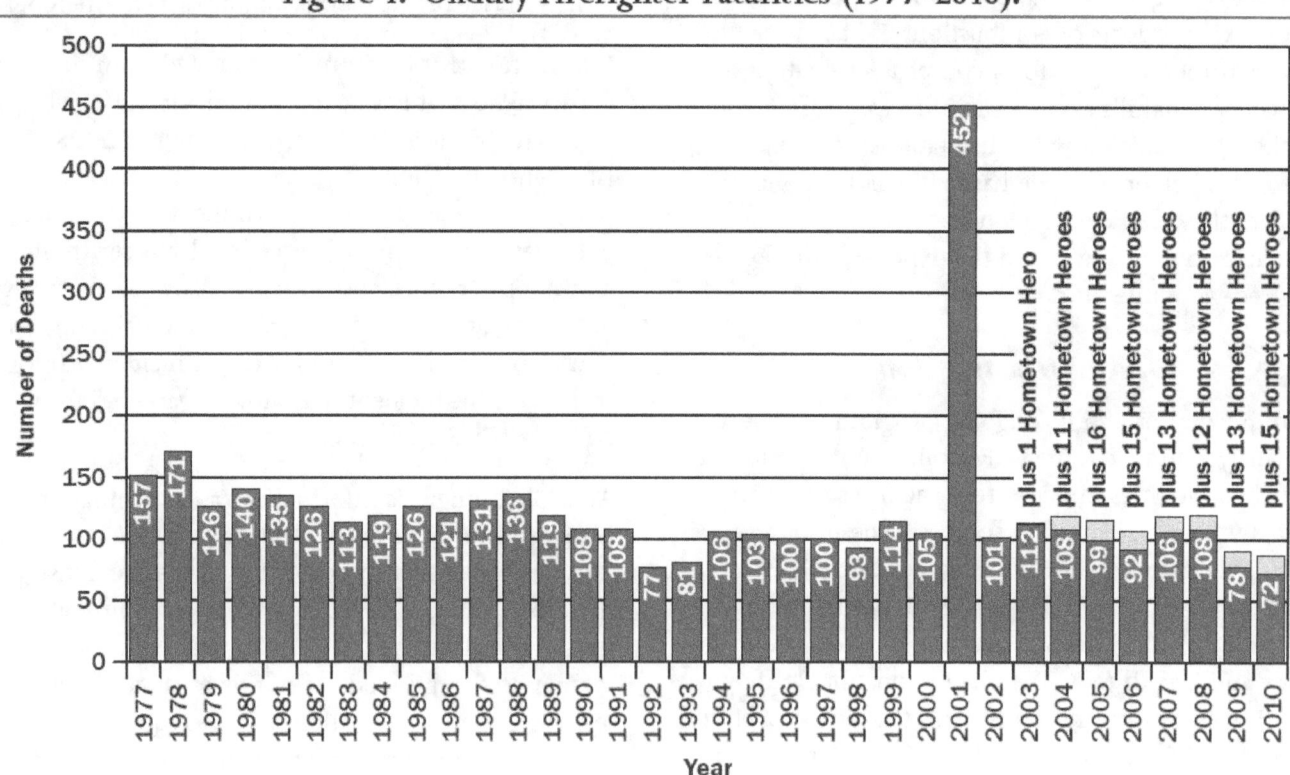

Figure 2. Firefighter Fatalities per 100,000 Fires.

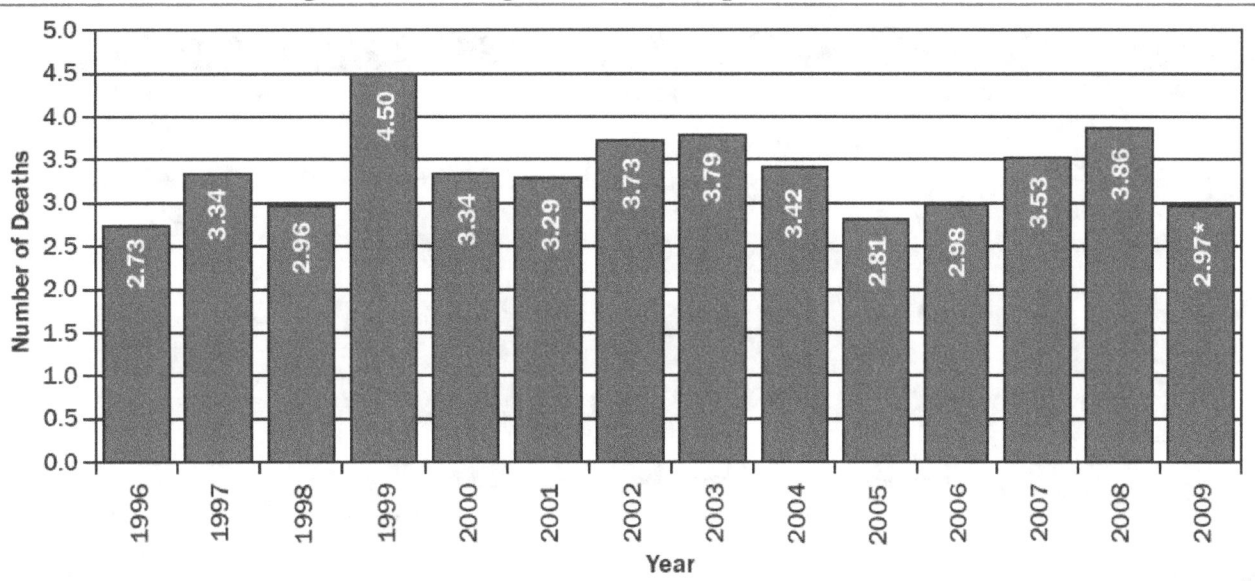

*2010 ratio will be included in the 2011 report.

Career, Volunteer, and Wildland Agency Deaths

In 2010, firefighter fatalities included 56 volunteer firefighters, 28 career firefighters, and three part-time or full-time members of wildland or wildland contract fire agencies (Figure 3).

Figure 3. Career, Volunteer, and Wildland Agency Deaths (2010).

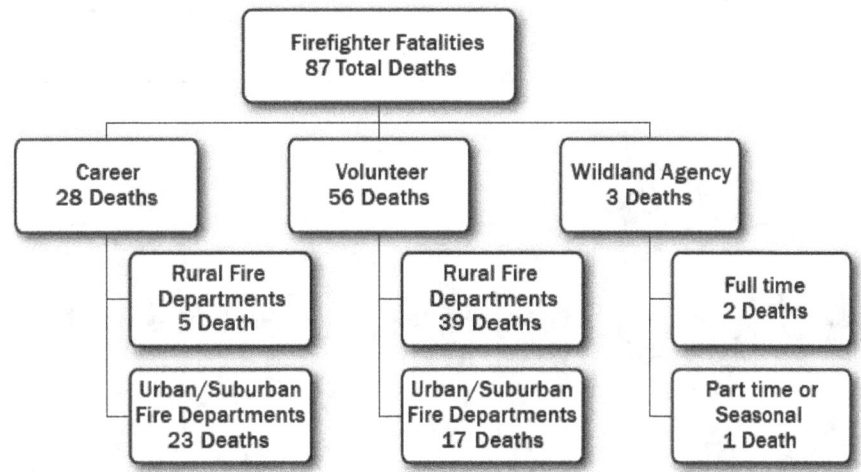

Gender

All of the firefighters who died while on duty in 2010 were male. This is the first year since 1998 where no female firefighters were killed while on duty. During the previous decade, 2000–2009, there were an average of four female, and not including WTC, 106 male onduty firefighter deaths per year.

Multiple Firefighter Fatality Incidents

The 87 deaths in 2010 resulted from a total of 83 fatal incidents. There were four firefighter fatality incidents where two firefighters were killed in each in 2010, claiming a total of eight firefighters.

Table 1. Multiple Firefighter Fatality Incidents

Year	Number of Incidents	Total Number of Deaths
2010	4	8
2009	6	13
2008	5	18
2007	7	21
2006	6	17
2005	4	10
2004	3	6
2003	7	20
2002	9	25
2001	8	362
2001 w/o WTC	7	18

Wildland Firefighting Deaths

In 2010, 11 firefighters were killed during activities involving brush, grass, or wildland firefighting. This total includes part-time and seasonal wildland firefighters, full-time wildland firefighters, and municipal or volunteer firefighters whose deaths are related to a wildland fire (Figure 4).

Figure 4. Firefighter Fatalities Related to Wildland Firefighting (2001–2010).

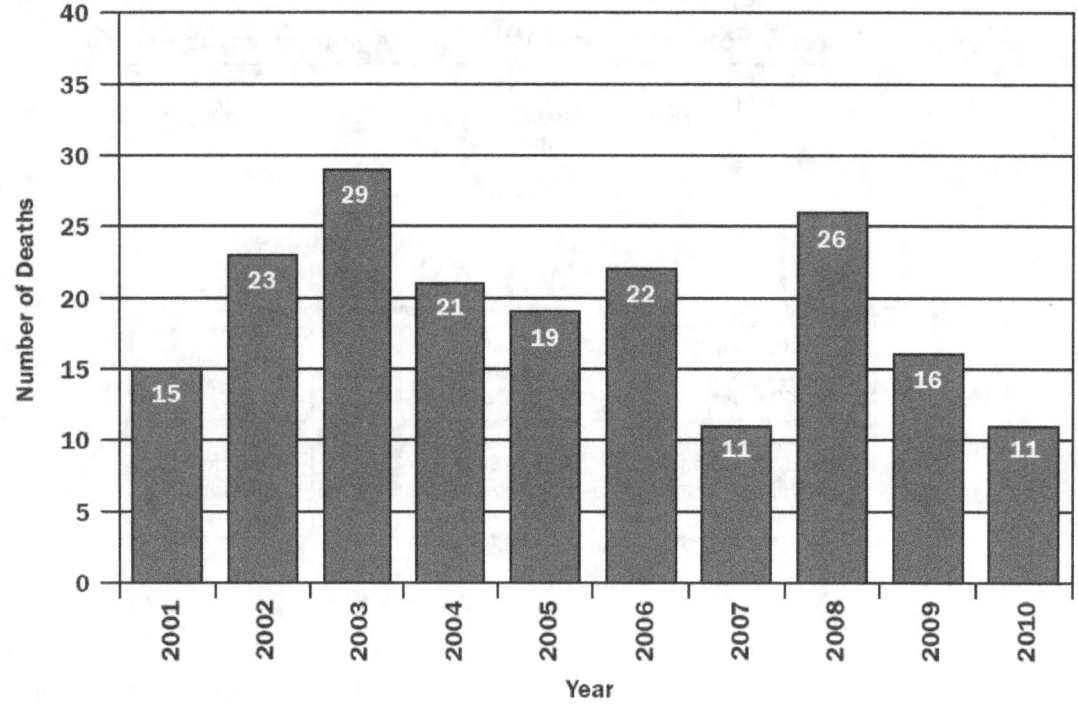

Table 2. Wildland Firefighting Aircraft Deaths

Year	Total Number of Deaths	Number of Fatal Incidents
2010	0	0
2009	5	3
2008	16	4
2007	1	1
2006	8	3
2005	6	2
2004	3	3
2003	7	4
2002	6	3
2001	6	3

In 2010, there were no multiple firefighter fatality incidents related to wildland firefighting and, for the first time since 1999, no wildland firefighter deaths related to aircraft.

Table 3. Firefighter Deaths Associated with Wildland Firefighting

Year	Total Number of Deaths	Number of Fatal Incidents	Number of Firefighters Killed in Multiple-Death Incidents
2010	11	11	0
2009	16	13	5
2008	26	15	14
2007	11	11	0
2006	22	13	13
2005	19	15	6
2004	21	21	0
2003	30	22	10
2002	23	14	13
2001	15	9	9

Activities related to emergency incidents resulted in the deaths of 48 firefighters in 2010 (Figure 5). This includes all firefighters who died responding to an emergency or at an emergency scene, returning from an emergency incident, and during other emergency-related activities. Nonemergency activities accounted for 39 fatalities. Nonemergency duties include training, administrative activities, performing other functions that are not related to an emergency incident, and postincident fatalities where the firefighter does not experience the illness or injury during the emergency.

Figure 5. Firefighter Deaths by Type of Duty (2010).

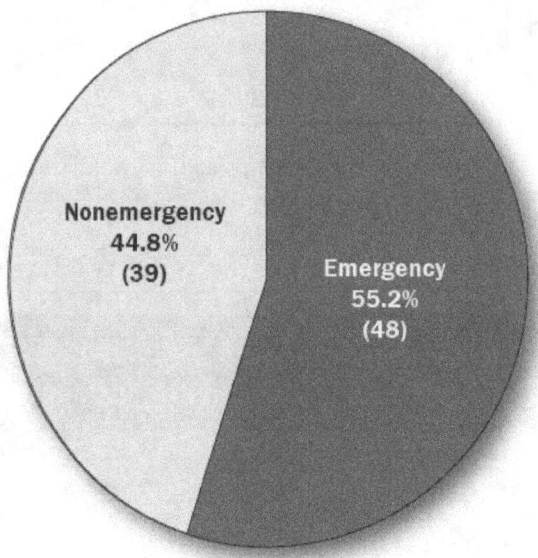

A multiyear historical perspective relating to the percentage of firefighter deaths that occurred during emergency duty is presented in Table 4.

Table 4. Emergency Duty Firefighter Deaths

Year	Percentage of All Deaths	Percentage of All Deaths Without Hometown Heroes
2010	55.2	66.7
2009	63.3	82.2
2008	63.5	70.0
2007	64.4	72.4
2006	57.5	66.3
2005	52.1	60.6
2004	68.9	75.9
2003	69.0	69.6
2002	73.0	N/A
2001	65.0	N/A
2001 with WTC	92.0	N/A
2000	71.0	N/A
1999	87.0	N/A
1998	77.0	N/A

The number of deaths by type of duty being performed in 2010 is shown in Table 5 and presented graphically in Figure 6. As has been the case for most years, fireground duties are the most common type of duty for firefighters killed while on duty.

Table 5. Firefighter Deaths by Type of Duty (2010)

Type of Duty	Number of Deaths
Fireground Operations	22
Other Onduty Deaths	18
Responding	15
Returning	1
Training	12
Onscene Nonfire Emergencies	4
After	15
Total	87

Figure 6. Firefighter Deaths by Type of Duty (2010).

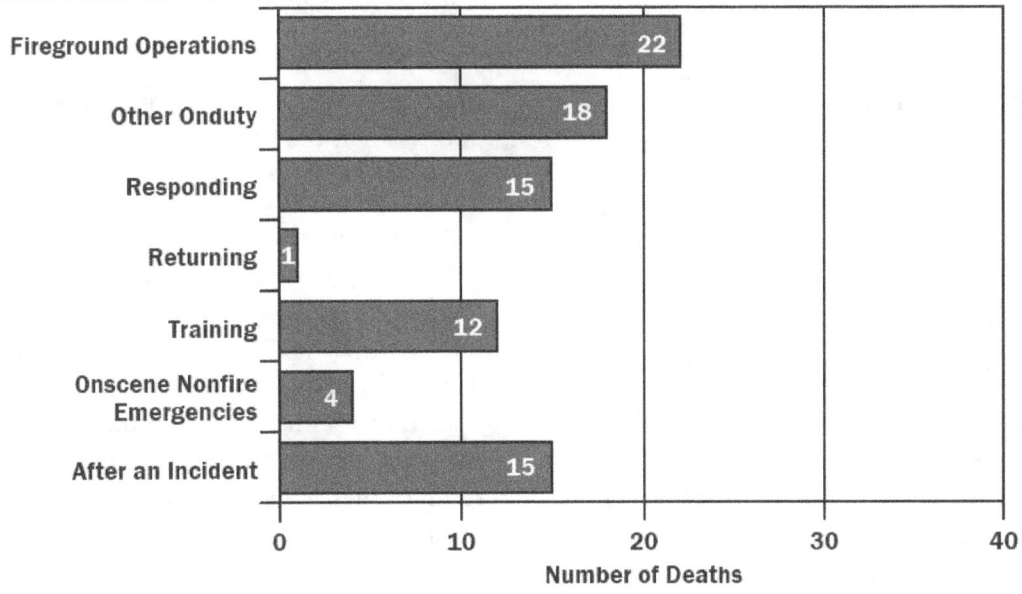

Fireground Operations

Of the 22 firefighters killed during fireground operations in 2010, 14 were at the scene of a structure fire, and 8 others were at the scene of a wildland or outside fire. The average age of the firefighters killed during fireground operations was 45, half of whom were from career fire departments, just under half (10) from volunteer fire departments, and one firefighter from a wildland agency.

Type of Fireground Activity

Table 6 shows the types of fireground activities in which firefighters were engaged at the time they sustained their fatal injuries or illnesses. This total includes all firefighting duties, such as wildland firefighting and structural firefighting.

Table 6. Type of Activity (2010)

Water Supply	1
Pump Ops	1
Search & Rescue	3
Advance Hoselines	10
Other	2
Unknown	4
Incident Command	1

Fixed Property Use for Structural Firefighting Deaths

There were 14 fatalities in 2010 where firefighters became ill or injured while on the scene of a structure fire. Table 7 shows the distribution of these deaths by fixed property use.

Table 7. Structural Firefighting Deaths by Fixed Property Use in 2010

Residential	8
Commercial	6

Responding/Returning

Sixteen firefighters died while responding to or returning from 15 emergency incidents in 2010. Fifteen of the firefighters died while responding to incidents and one while returning from an incident.

Nine of the firefighters killed while responding to incidents died from heart attacks (eight) or a stroke (one).

Six of the firefighters who died while responding to incidents were killed by trauma caused by motor vehicle collisions, including three in privately-owned vehicles (POVs) and three in fire department apparatus. Of the POV crashes taking a firefighter's life, one occurred on an all-terrain vehicle (ATV) that struck a deer. No helmet and alcohol, the only 2010 incident where alcohol was reported to be involved, were contributing factors in the death. In the two other separate POV incidents, seat restraints were present in both vehicles. One firefighter was wearing them and one was not; the latter fully ejected from the vehicle.

Of the three firefighters killed in fire department apparatus involved in motor vehicle collisions, two were not wearing seat restraints and fully ejected when their engine was struck at an intersection by a sport utility vehicle (SUV). The third firefighter killed, who was belted in and not ejected from the vehicle, was operating a semitrailer truck hauling a D5 Caterpillar plow that did not successfully negotiate a curve. As a result of the accident, the firefighter became trapped in the cab of the truck and was pronounced dead at the scene.

The one firefighter killed while returning from an incident was crushed between a moving tanker and a parked apparatus as the tanker backed into the station's apparatus bay.

Table 8. Firefighter Deaths While Responding to or Returning from an Incident

Year	Number of Firefighter Deaths
2010	16
2009	15
2008	24
2007	26
2006	15
2005	22
2004	23
2003	36
2002	13
2001	23

Training

In 2010, 12 firefighters died while engaged in training activities. Seven of the deaths were due to heart attacks, two from cerebrovascular accidents (CVA), and one from a documented case of necrotizing fasciitis (commonly known as flesh eating disease). An additional two firefighters were killed in a POV while they were returning from a fire certification class at a local community college. These firefighters were apparently involved in street racing with firefighters in another vehicle.

Table 9. Firefighter Fatalities While Engaged in Training

Year	Number of Firefighter Deaths
2010	12
2009	10
2008	12
2007	11
2006	9
2005	14
2004	13
2003	12
2002	11
2001	14

Nonfire Emergencies

In 2010, there were two fire police and two firefighter fatalities where the type of emergency duty was not related to a fire. One fire police was struck and killed by a vehicle while performing scene safety duties at the scene of a motor vehicle accident. Another fire police suffered a heart attack after establishing a medivac helispot to evacuate a patient from an emergency medical services (EMS) incident. The two firefighters both died from asphyxiation in separate incidents. The first firefighter was part of a three-person crew in a boat and actively engaged in rescue operations when the boat was captured by strong currents, struck a bridge, and capsized. The second firefighter died as a result of oxygen deprivation in a sewer system manhole while he attempted to rescue a member of the town's Department of Public Works (DPW) who had lost consciousness after descending the interior ladder and falling to the bottom.

After the Incident

In 2010, 15 firefighters died after the conclusion of their onduty activity. All 15 deaths were due to heart attacks.

CAUSE OF FATAL INJURY

The term "cause of injury" refers to the action, lack of action, or circumstances that directly resulted in the fatal injury. The term "nature of injury" refers to the medical cause of the fatal injury or illness which is often referred to as the physiological cause of death. A fatal injury is usually the result of a chain of events, the first of which is recorded as the cause.

In 2010, one firefighter was reported to have committed suicide while on duty.

Figure 7 shows the distribution of deaths by cause of fatal injury or illness in 2010.

Figure 7. Fatalities by Cause of Fatal Injury (2010).

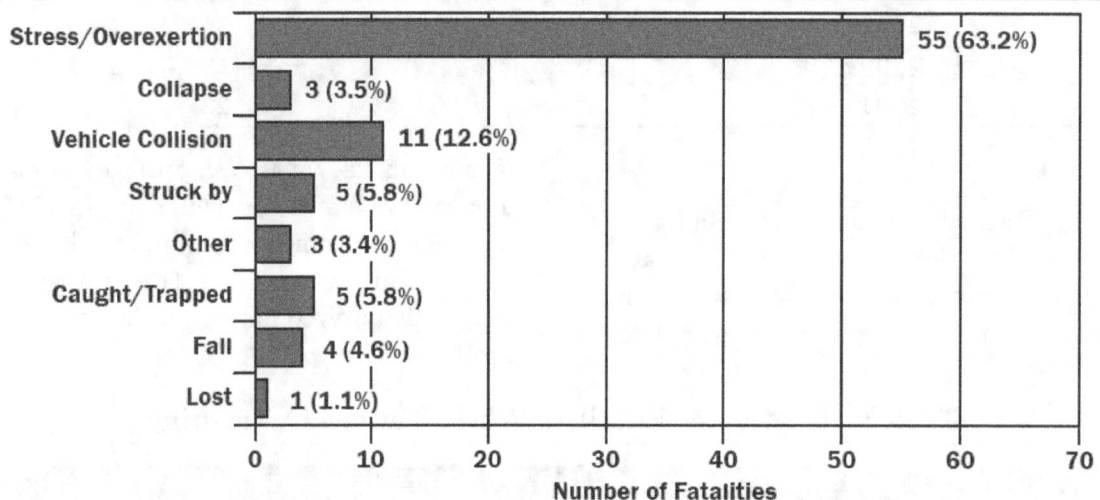

Stress or Overexertion

Firefighting is extremely strenuous physical work and is likely one of the most physically demanding activities that the human body performs.

Stress or overexertion is a general category that includes all firefighter deaths that are cardiac or cerebrovascular in nature such as heart attacks, strokes, and other events such as extreme climatic thermal exposure. Classification of a firefighter fatality in this cause of fatal injury category does not necessarily indicate that a firefighter was in poor physical condition.

Fifty-five firefighters died in 2010 as a result of stress/overexertion:

- Fifty firefighters died due to heart attacks.

- Five firefighters died due to CVAs.

Table 10. Deaths Caused by Stress or Overexertion

Year	Number	Percent of Fatalities
2010	55	63.2
2009	50	55.5
2008	52	44.0
2007	55	46.6
2006	54	50.9
2005	62	53.9
2004	66	56.4
2003	51	45.9**
2002	38	38.0
2001	43	40.9*

*Does not include the firefighter deaths of September 11, 2001, in New York City.
**Includes Hometown Heroes, one in December 2003, and an average of 13.4 for the years 2004–2010.

Vehicle Crashes

Eleven firefighters died in 2010 as the result of vehicle crashes, down 31 percent from vehicle crashes in 2009 and almost 61 percent from 2008. In 2010, there were no firefighter deaths that involved aircraft.

In addition to the responding/returning from an incident and training-related vehicle crash fatalities earlier in this report, there were two additional vehicle crashes that killed firefighters in 2010:

- While on scene of a mutual-aid wildland fire call, a nontraffic vehicle (snowcat) ascended a hill, lost traction, and began to tumble down the slope, killing the operator as it rolled more than three times. Another firefighter was ejected from the snowcat but survived the accident.

- A wildland crew carrier, with as many as 12 firefighters in the vehicle, was struck by an oncoming SUV that had crossed over the center line of the highway. Several firefighters were ejected in the crash and a number received severe injuries, including one who died from his injuries.

Figure 8. Firefighter Fatalities in Vehicle Collisions.

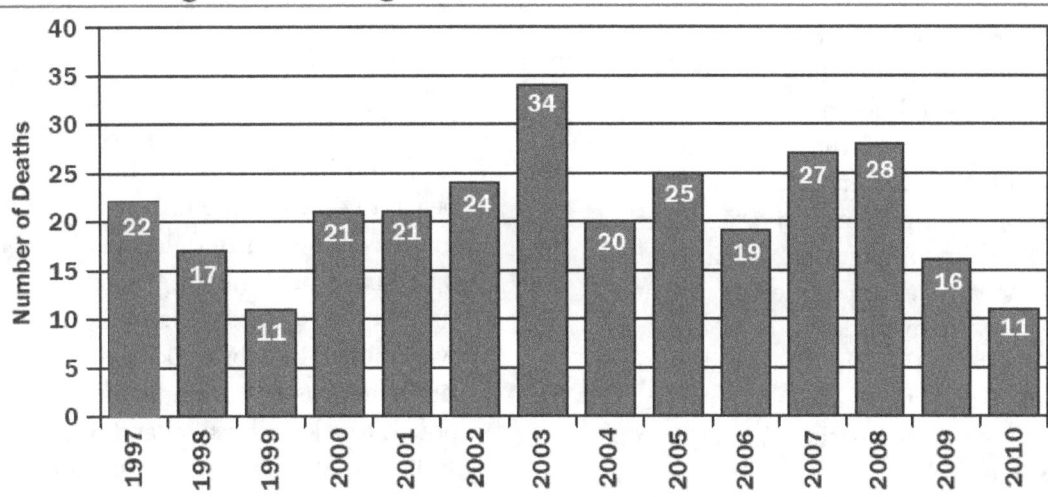

Lost or Disoriented

One firefighter died in 2010 when he became lost or disoriented inside of a burning residential structure while searching for an elderly couple and after having rescued their dog, was reported to still be inside of a burning residential structure. Rapid Intervention Team (RIT) crews were assigned and searched the structure for the firefighter. After approximately 10 minutes of searching by multiple crews, he was located in a small room behind a closed door. Investigation revealed that the firefighter had become ill and vomited into his self-contained breathing apparatus (SCBA) facepiece. When he was found, he was lying on his back without his helmet, gloves, and facepiece. His death was caused by smoke inhalation.

Caught or Trapped

Five firefighters were killed in 2010 in four separate incidents when they became caught or trapped. This classification covers firefighters trapped in wildland and structural fires who were unable to escape due to

rapid fire progression and the byproducts of smoke, heat, toxic gases, and flame. This classification also includes firefighters who drowned and those who were trapped and crushed.

- Firefighters conducting search activities in a residential structure observed fire advancement and yelled to the hoseline crew to evacuate. Once firefighters exited, they found that two firefighters were still in the structure. Fire conditions inside had dramatically changed. One firefighter was able to make it to within a few feet of the door as he was pulled from the structure by other firefighters. Firefighters entered the structure with a hoseline to search for the other missing firefighter. He was found wrapped up in the ruptured 2-1/2 inch handline and, although wearing his facepiece when he entered the structure, was not wearing his facepiece, hood, or helmet. The firefighter was removed from the structure, treated by other firefighters, and transported to the hospital by ambulance, but did not survive.

- One firefighter and his engine company arrived approximately 20 minutes after dispatch and found a working fire in a large egg production and processing facility. Two firefighters advanced an attack line into the structure but became separated from the hoseline and then from each other in the interior. One firefighter was able to find a metal exterior wall and banged on it until firefighters on the exterior were able to cut open the wall to allow his escape. The fire progressed and attempts to locate the second firefighter were unsuccessful. He died of thermal injuries and smoke inhalation.

- Upon their arrival at the scene, firefighters found an active fire on the second floor of a residential structure. Two firefighters were assigned to go to the third floor of the building to look for fire extension and to search for victims. When both firefighters, who were wearing full structural protective clothing, including SCBAs, arrived on the third floor, one firefighter reported stable conditions by radio to the Incident Commander (IC) and requested that a hoseline be brought to the floor. A few moments later, the IC observed smoke coming from the third floor and attempted to contact the firefighter by radio to no avail. The IC ordered an evacuation of the building by all firefighters in order to account for everyone on the scene. In the process of the evacuation, firefighters were observed on the third floor in distress. The RIT was deployed to the third floor for rescue. Both firefighters were found unconscious by firefighters and brought to the exterior of the building. Firefighters and emergency responders immediately began to provide medical treatment for both firefighters and they were transported to the hospital but did not survive their injuries. Autopsies and toxicological studies showed that both died of smoke inhalation suffered as they fought the structure fire.

- Firefighters were requested to assist Department of Public Works (DPW) employees with access to a sewer system manhole behind their fire station. A DPW employee entered the manhole and lost consciousness as he descended the interior ladder and fell to the bottom. The DPW foreperson requested assistance from firefighters. The fire chief asked dispatch for EMS assistance and also started a fire

department incident. As an atmospheric meter from the fire department was prepared for use, a firefighter entered the opening. He was not wearing a SCBA or harness. When the firefighter was halfway down the ladder, he lost consciousness and fell to the bottom. Additional resources were called to the scene and both the firefighter and the DPW employee were removed from the opening; both died of asphyxiation due to an oxygen deficient atmosphere.

Collapse

Three firefighters died in 2010 as a result of structural collapses in two separate commercial property fire incidents.

- Firefighters found a working fire in an office building and initiated an interior attack. Firefighters found fire in the attic of the structure and began to evacuate the interior. Approximately 30 minutes into the fire fight, two firefighters were operating a handline on the exterior of the structure when a collapse occurred. One firefighter was trapped in debris and was removed by other firefighters using hydraulic tools. He was transported to a local hospital and was treated for crush and thermal injuries. He was transported to a regional hospital and then to a care facility where he died almost 5 months later from injuries sustained.

- A small, intentionally-set rubbish fire had been quickly extinguished with tank water and extension into the trusses was ruled out in a vacant commercial building. The east wall (sector 4 side) of the structure had three large wire glass metal framed industrial-type windows. Firefighters began removing these windows to provide a quick means of egress for firefighters working within the structure. As the southern-most window on the east wall was removed, it is theorized that the entire east wall failed. The east wall was a bearing wall which supported the weight of two heavy timber bowstring trusses. The failure of the east wall caused total collapse of the bowstring truss roof and pushed out the sector 3 wall. Sixteen firefighters became trapped and injured in the collapse. Two firefighters passed away from injuries sustained in this incident.

Struck by Object

Being struck by an object was the cause of five fatal firefighter injuries in 2010.

- One fire police was working scene safety and traffic control services for a mutual-aid motor vehicle crash with entrapment and wires down. He parked his vehicle on the side of the road with its warning lights in operation and set cones and a flare in the roadway to warn approaching drivers. A civilian vehicle disregarded the road closure, drove over cones, and struck the fire police officer. He was transported to the hospital but was pronounced dead upon arrival.

- One firefighter was clearing a fire break for containment at the base of a bluff when a burning snag broke loose on top and rolled downhill over a small bluff, striking him from behind. The firefighter sustained a serious head injury, fractured hip, bruises, and second-degree burns on his calves. The impact left him unconscious and with serious injuries, including the burns from which he did not recover.

- One firefighter was staffing a brush truck while fighting a wildland fire when the water tank on the brush truck failed under pressure and the firefighter was struck in the chest by a piece of the tank. The firefighter was killed as a result of trauma and was pronounced dead at the scene.

- A firefighter/mechanic was making steering system repairs to a brush truck inside of the fire department's maintenance building. As the mechanic turned the key to unlock the steering column, the brush truck accidentally lunged forward. At the same time the brush truck lunged forward, two firefighters were walking in front of the vehicle. One of the firefighters was knocked clear but the other was crushed between the brush truck and a wall. He received a head injury and died instantly from his injuries.

- Two engine companies were dispatched to a fire in the median of a local highway. Upon arrival, engine personnel positioned their apparatus in a shielding position to protect firefighters from traffic. As firefighters were extinguishing the fire, a vehicle crash occurred on the highway behind the apparatus. The crash caused one of the vehicles to slide between the guardrail and one of the engines. The vehicle entered the fire scene and struck two firefighters. Both firefighters were transported to the hospital but one of them died due to traumatic injuries.

Fall

Four firefighters died in 2010 as the result of falls.

- After a firefighter had closed and latched the doors at the top of a burning grain silo, he fell 20–30 feet when an explosion occurred inside of the silo. The firefighter was transported to the hospital by ambulance but was pronounced dead at the hospital due to traumatic injuries.

- A firefighter working in-station duties experienced a fall. Firefighters treated him for a cut on his head and recommended that he go to the hospital to be checked. Later that night, they checked on their fellow firefighter in the hospital and found him in better condition than before. The firefighter, however, later took a turn for the worse and died from the head injury received in the fall at the fire station.

- A firefighter and his ladder company were dispatched to a structural fire in a four-story commercial and residential structure. Upon their arrival, firefighters observed smoke and sparks coming from a cooking exhaust fan chute. The firefighter ascended a ladder attached to the side of the building to gain access to the roof. He was wearing full structural firefighting protective clothing, including SCBA. He also carried a water fire extinguisher with him as he climbed. When the firefighter reached the roof, he lost his grip and fell 53 feet to the ground. The firefighter was severely injured in the fall and was treated by firefighters and transported to the hospital. He was pronounced dead at the hospital due to multiple injuries.

- Members of a fire department were working on the construction of a new fire station to replace one that had been destroyed by fire. One firefighter died from injuries received on the worksite when he fell backwards and struck his head on concrete.

Other

Three firefighters died in 2010 of a cause that is not categorized above.

- One firefighter was struck by an Amtrak train. State police, along with Amtrak police and the city, investigated the circumstances surrounding the death. No foul play was suspected.

- One firefighter was a part of a three-person crew in a boat and actively engaged in rescue operations when it was captured by strong currents, struck a bridge, and capsized.

- During the course of a training exercise, a firefighter struck his elbow on a portion of the body of an engine as he removed SCBA cylinders for use in the training. He noted his injury to other firefighters but thought that it was insignificant. The firefighter signed out of training but approximately 6 hours later awoke and began vomiting. He was also experiencing pain is his injured elbow. An ambulance was called. Arriving EMS responders treated the incident as a cardiac event based on symptoms and transported him to the hospital. The firefighter was diagnosed as suffering from necrotizing fasciitis (commonly known as flesh eating disease). He underwent multiple surgeries to attempt to control the spread of the disease and was transferred to a burn unit for treatment. The firefighter underwent a total of 18 operations during 79 days in the hospital. Although his condition seemed to improve for a time, he died as a result of complications related to the injury.

NATURE OF FATAL INJURY

Figure 9 shows the distribution of the 87 firefighter deaths that occurred in 2010 by the medical nature of the fatal injury or illness. For heart attacks, Figure 10 shows the type of duty involved.

The National Institute for Occupational Safety and Health's (NIOSH's) Fire Fighter Fatality Investigation and Prevention Program (http://www.cdc.gov/niosh/fire/) is the leading resource in the world for investigations and focused reports on firefighter fatality incidents in the United States of America.

Figure 9. Fatalities by Nature of Fatal Injury (2010).

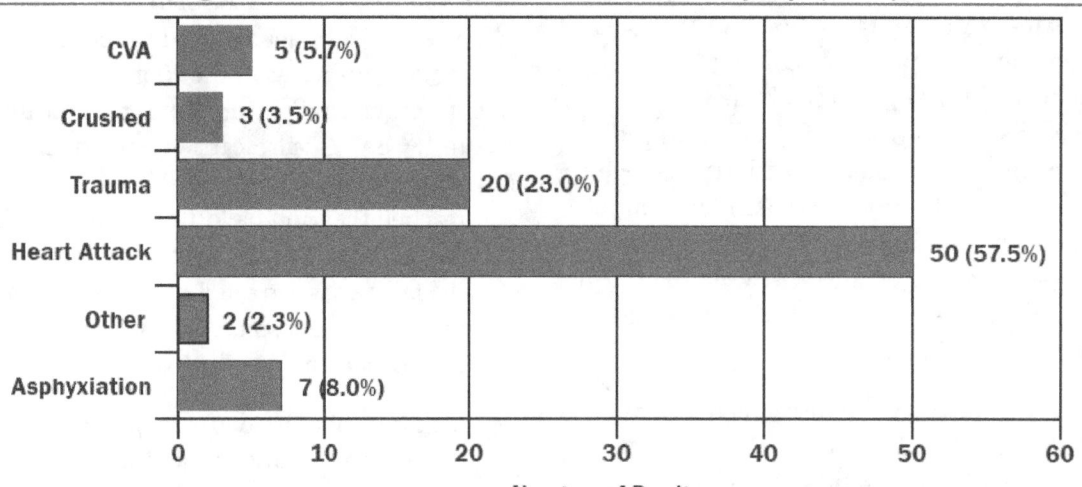

Figure 10. Heart Attacks by Type of Duty (2010).

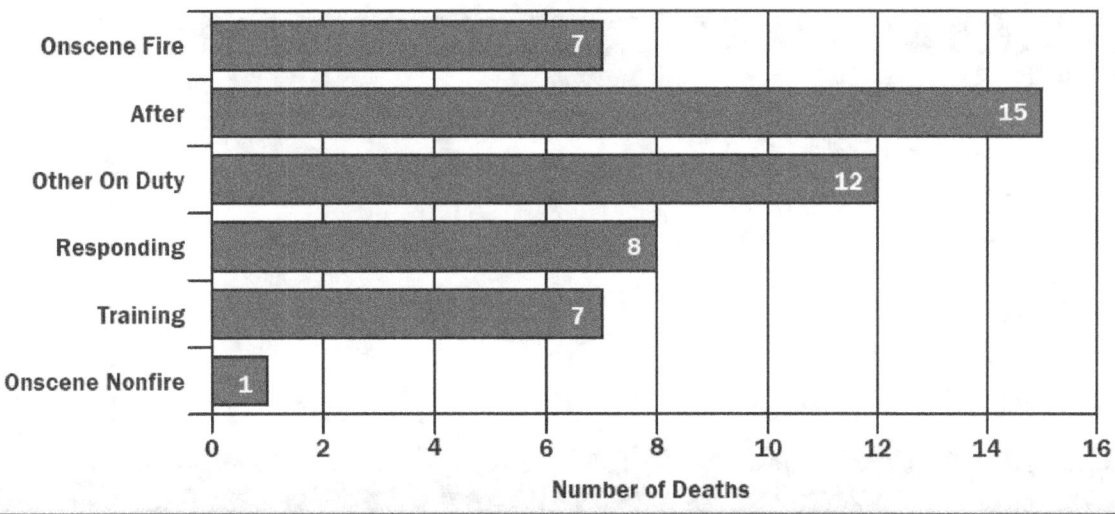

FIREFIGHTER AGES

Figure 11 shows the percentage distribution of fire-fighter deaths by age and nature of the fatal injury. Table 11 provides a count of firefighter fatalities by age and the nature of the fatal injury.

Younger firefighters were more likely to have died as a result of traumatic injuries, such as injuries from an apparatus accident or becoming caught or trapped during firefighting operations. Stress-related deaths are rare below the 31 to 35 years of age category and, when they occur, often include underlying medical conditions.

Figure 11. Fatalities by Age and Nature (2010).

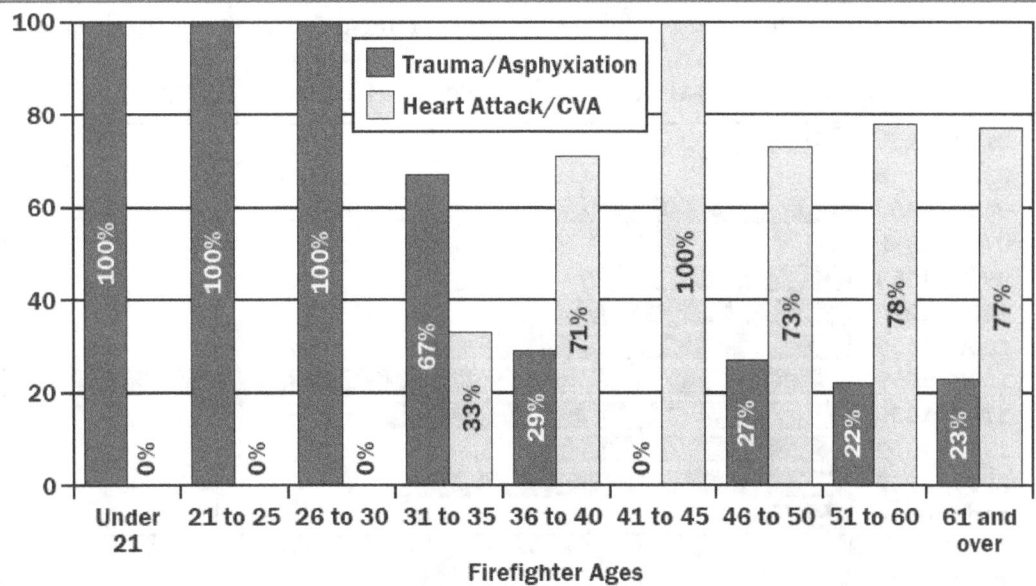

Table 11. Firefighter Ages and Nature of Fatal Injury (2010)

Age Range	Heart Attack/CVA/Other	Trauma/Asphyxiation Total
under 21	0	2
21 to 25	0	5
26 to 30	0	4
31 to 35	2	4
36 to 40	5	2
41 to 45	4	0
46 to 50	8	3
51 to 60	18	4
61 and over	20	6

In 2010, there were no teenage firefighters killed while on duty in the United States for the first time since 1993. The two youngest firefighters to die in 2010 were both age 20. Both firefighters were killed in separate incidents involving POV crashes. One firefighter was responding to a fatal motor vehicle accident on a lo cal highway. As he approached the scene, the firefighter, who was not wearing seat restraints, lost control of his pickup truck, struck a guardrail, rolled, and was ejected. The second 20 year old firefighter died while returning from training and riding as a passenger in a car that crashed traveling at high speed while involved in street racing against other firefighters in another vehicle. The accident also claimed the life of another firefighter (driver) and a private citizen riding in another vehicle in the opposite lane of travel who died from injuries nearly 2 weeks later.

The oldest firefighter killed on duty in 2010 was 86. Operating as fire police, he suffered a heart attack after establishing a medivac helispot to evacuate a patient from an EMS incident.

DEATHS BY TIME OF INJURY

The distribution of 2010 firefighter deaths according to the time of day when the fatal injury occurred is illustrated in Figure 12. The time of fatal injury for four firefighters was either unknown or not reported.

Figure 12. Fatalities by Time of Fatal Injury (2010).

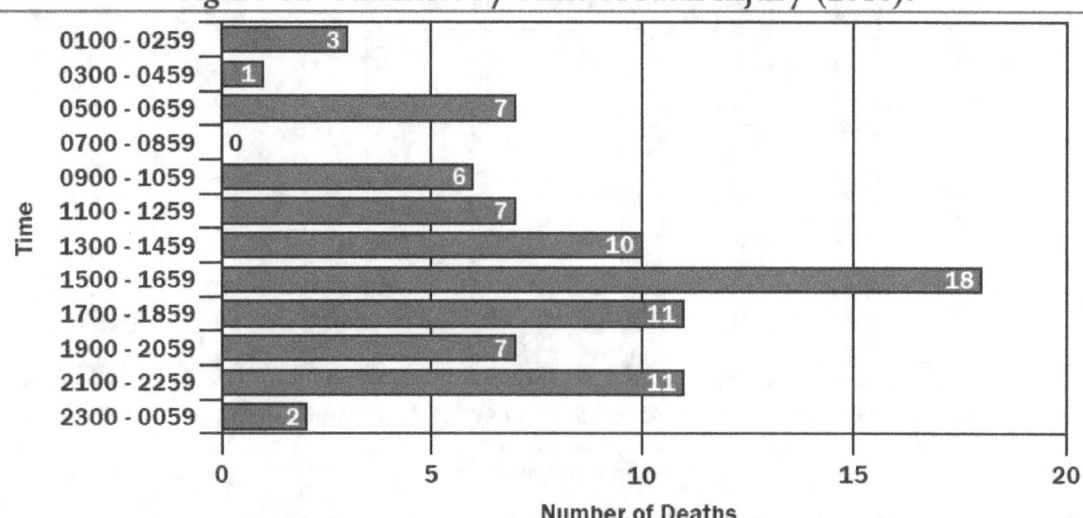

FIREFIGHTER FATALITY INCIDENTS BY MONTH OF YEAR

Figure 13 illustrates the 2010 firefighter fatalities by month of the year.

Figure 13. Deaths by Month of Year (2010).

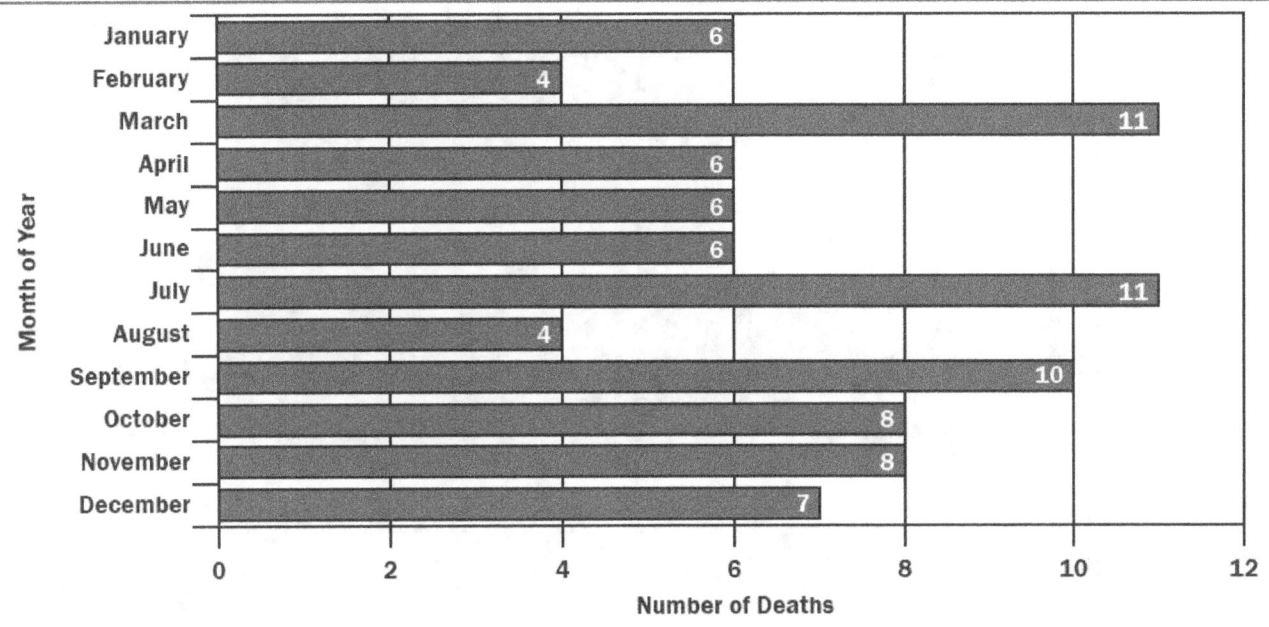

STATE AND REGION

The distribution of firefighter deaths in 2010 by State is shown in Table 12. Firefighters based in 31 States died in 2010.

The highest number of firefighter deaths, based on the location of the fire service organization in 2010, occurred in Illinois with nine deaths. New York and Ohio had the next highest totals of firefighter fatalities in 2010, with eight firefighter deaths each, respectively.

Table 12. Firefighter Fatalities by State by Location of Fire Service* (2010)

State	Fatalities	Percentage
OH	8	9
AR	3	3
NY	8	9
MS	1	1
KY	3	3
ID	1	1
KS	5	5
PA	7	8
DE	1	1
ME	1	1
CT	4	4
WV	1	1

State	Fatalities	Percentage
NC	3	3
AZ	2	2
IL	9	10
MO	2	2
IA	1	1
OK	1	1
WI	1	1
NJ	2	2
GA	2	2
WA	1	1
SC	3	3
TX	2	2
VT	1	1
MA	2	2
VA	3	3
CA	3	3
IN	4	4
TN	1	1
MI	1	1
OH	8	9

*This list attributes the deaths according to the State in which the fire department or unit is based, as opposed to the State in which the death occurred. They are listed by those States for statistical purposes and for the National Fallen Firefighters Memorial at the National Emergency Training Center (NETC). Due to rounding, percentage totals may not add up to 100.

Figure 14. Firefighter Fatalities by Region (2010).

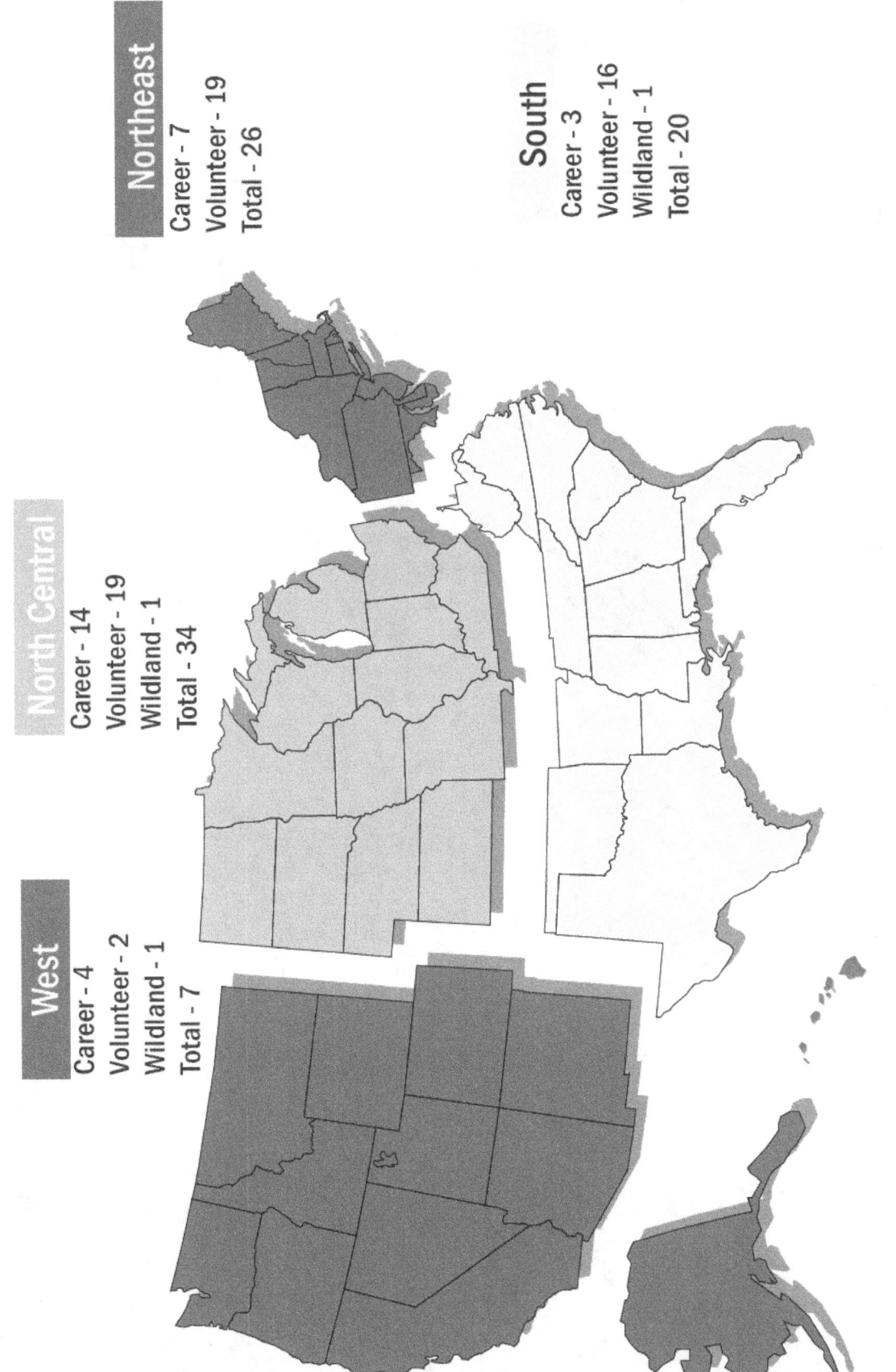

Northeast
Career - 7
Volunteer - 19
Total - 26

South
Career - 3
Volunteer - 16
Wildland - 1
Total - 20

North Central
Career - 14
Volunteer - 19
Wildland - 1
Total - 34

West
Career - 4
Volunteer - 2
Wildland - 1
Total - 7

Figure 15. Onduty Firefighter Fatalities 2010 by Fire Department Location.

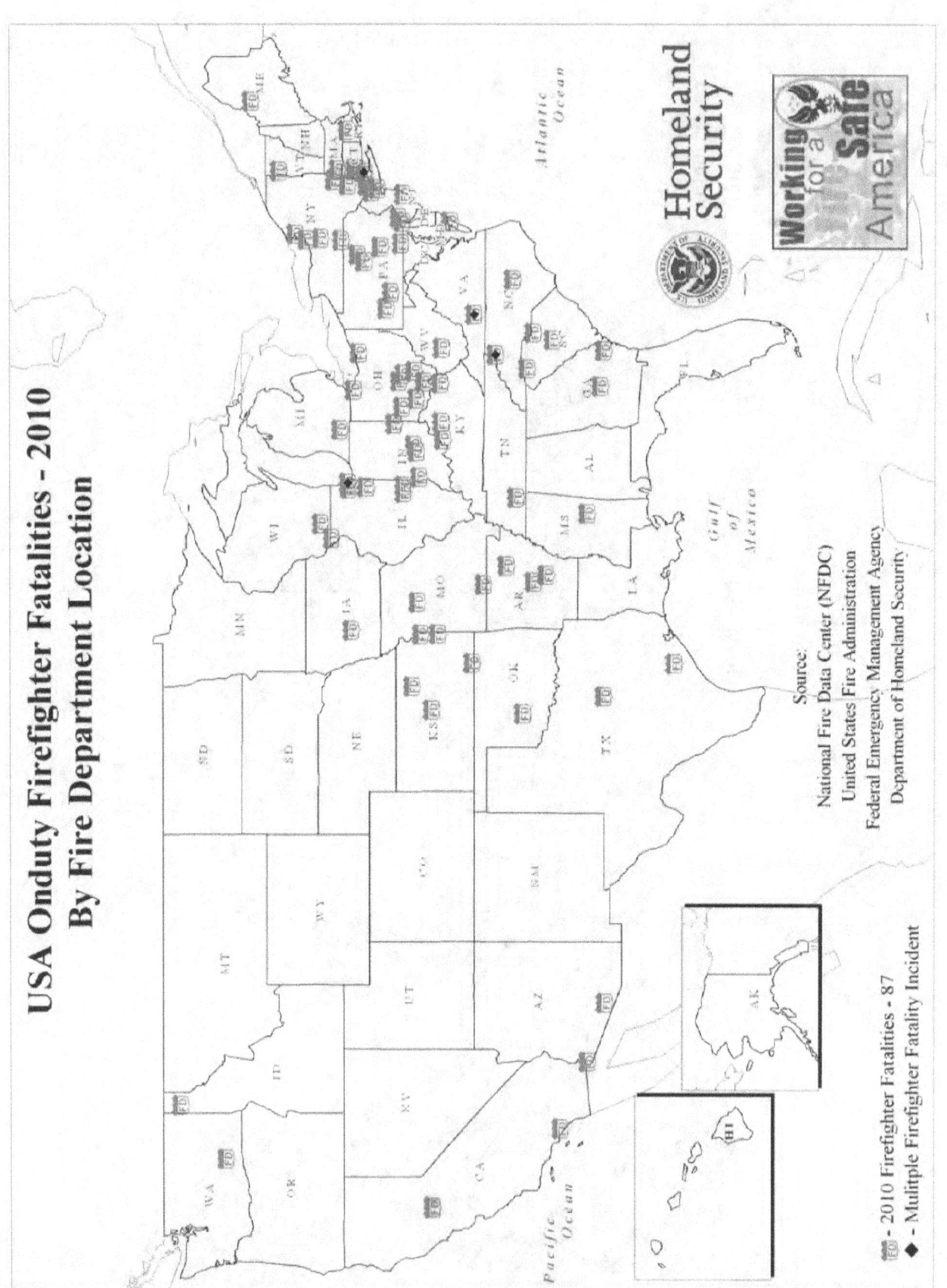

USA Onduty Firefighter Fatalities – 2010
By Fire Department Location

Source:
National Fire Data Center (NFDC)
United States Fire Administration
Federal Emergency Management Agency
Department of Homeland Security

Homeland Security

Working for a Safe America

🏠 - 2010 Firefighter Fatalities - 87
◆ - Mulitple Firefighter Fatality Incident

Figure 16. Onduty Firefighter Fatalities 2010 by Incident Location.

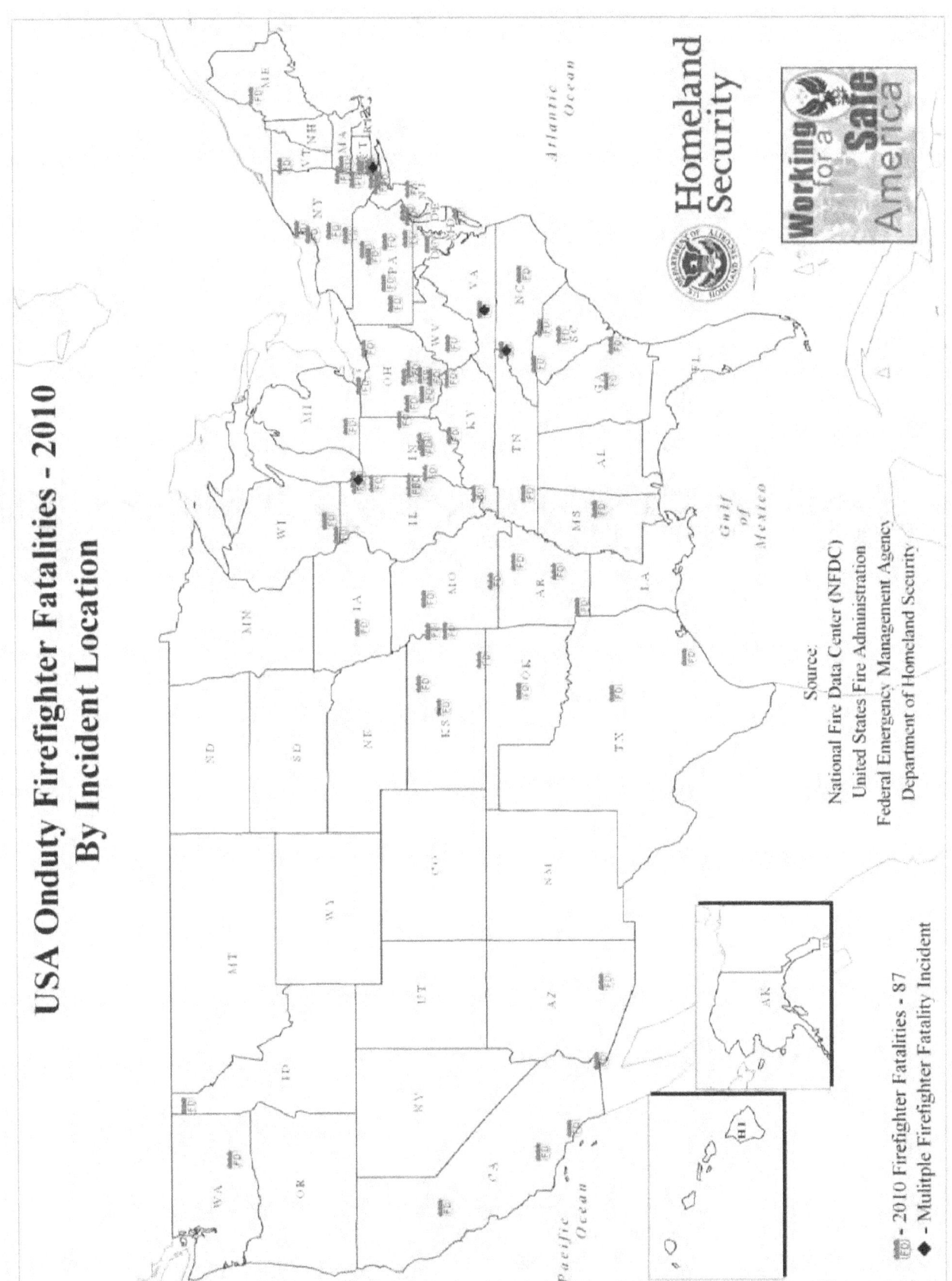

ANALYSIS OF URBAN/RURAL/SUBURBAN PATTERNS IN FIREFIGHTER FATALITIES

The U.S. Census Bureau defines "urban" as a place having a population of at least 2,500 or lying within a designated urban area. "Rural" is defined as any community that is not urban. "Suburban" is not a census term but may be taken to refer to any place, urban or rural, that lies within a metropolitan area defined by the Census Bureau, but not within one of the central cities of that metropolitan area.

Fire department areas of responsibility do not always conform to the boundaries used by the Census Bureau. For example, fire departments organized by counties or special fire protection districts may have both urban and rural coverage areas. In such cases, where it may not be possible to characterize the entire coverage area of the fire department as rural or urban, firefighter deaths were listed as urban or rural based on the particular community or location in which the fatality occurred.

The following patterns were found for 2010 firefighter fatalities. These statistics are based on answers from the fire departments and, when no data from the departments were available, the data were based upon population and area served as reported by the fire departments.

Table 13. Firefighter Deaths by Coverage Area Type (2010)

	Urban/Suburban	Rural	Total
Firefighter Deaths	40	47	87

In memory of all firefighters
who answered their last call in 2010

To their families and friends

To their service and sacrifice

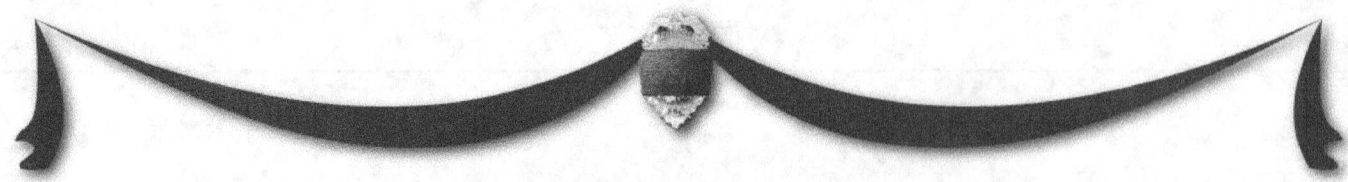

January 2, 2010–1827 hrs
Joseph Mack McCafferty, Lieutenant
Age 59, Career • Lancaster Fire Department, Ohio

Lieutenant McCafferty and the members of his fire department were dispatched at 1727 hours to a structure fire in a two-story wood-frame apartment building. On arrival, firefighters on the scene reported a working fire.

Lieutenant McCafferty was in command of the incident. At approximately 1827 hours, Lieutenant McCafferty told a firefighter/paramedic that he had a severe headache, back and neck pain, and that he felt nauseous. He was assisted to an ambulance on scene by other firefighters. In the ambulance, his condition worsened and he was transported to a local hospital.

Lieutenant McCafferty was treated at the local hospital and then transported to a regional hospital. He died on January 16, 2010. The cause of death was a cerebrovascular accident (CVA). Lieutenant McCafferty was the first line-of-duty death (LODD) for the Lancaster Fire Department in its 117-year history.

January 3, 2010–0515 hrs
David Allen Curlin, Lieutenant
Age 40, Career • Pine Bluff Fire and Emergency Services, Arkansas

Lieutenant Curlin and the members of his fire department were dispatched to a fire in a commercial occupancy at 0445 hours. Firefighters found a working fire in an office building and initiated an interior attack. Firefighters found fire in the attic of the structure and began to evacuate the interior.

Approximately 30 minutes into the fire fight, Lieutenant Curlin and another firefighter were operating a handline on the exterior of the structure when a collapse occurred. Lieutenant Curlin was trapped in debris and was removed by other firefighters using hydraulic tools. He was transported to a local hospital and was treated for crush and thermal injuries. He was transported to a regional hospital and then to a care facility. He died on May 22, 2010. Lieutenant Curlin's death was a result of the injuries he received on January 3, 2010.

January 13, 2010–0615 hrs
LeRoy Arthur Kemp, Firefighter/Past Chief
Age 81, Volunteer • Tioga Center Fire Department, New York

Firefighter Kemp was at home when his fire department was dispatched to an emergency medical incident in a neighboring town. Firefighter Kemp responded from home in his personal vehicle with his emergency light in operation.

As he responded, Firefighter Kemp came upon the scene of an earlier traffic crash involving a refuse vehicle. Firefighter Kemp was unable to stop in time and struck the rear of the refuse vehicle. He was killed instantly in the crash.

January 14, 2010–1315 hrs
Jerry W. Thompson, Firefighter
Age 55, Volunteer • Linwood Volunteer Fire Department, Mississippi

Firefighter Thompson was engaged in overhaul duties at the scene of a fire in a manufactured home when he suffered a heart attack. He was treated at the scene by firefighters and emergency medical services (EMS) responders and flown by helicopter to a hospital. He did not recover and died later that day.

For additional information regarding this incident, please refer to the National Institute for Occupational Safety and Health (NIOSH) Fire Fighter Fatality Investigation and Prevention Program Report F2010-29 (www.cdc.gov/niosh/fire/reports/face201029.html).

January 17, 2010–1740 hrs
Terry Leo Cannon, Major
Age 51, Volunteer • Buechel Fire Protection District, Kentucky

Major Cannon and the members of his fire department responded to a fire alarm activation in a residence. The incident was caused by steam from a shower and was concluded at 1340 hours. Later that afternoon, Major Cannon worked out at a local fitness facility. After completing a treadmill run, he collapsed at approximately 1740 hours. Staff from the fitness facility initiated cardiopulmonary resuscitation (CPR) and summoned help.

Firefighters from the Buechel Fire Protection District responded to the scene and treated Major Cannon. An automatic external defibrillator (AED) was applied and advanced life support (ALS) care was provided by emergency medical responders.

Continued on next page.

Major Cannon was transported to the hospital by ambulance, and was later pronounced dead. His death was caused by a heart attack.

For additional information regarding this incident, please refer to NIOSH Fire Fighter Fatality Investigation and Prevention Program Report F2010-08 (www.cdc.gov/niosh/fire/reports/face201008.html).

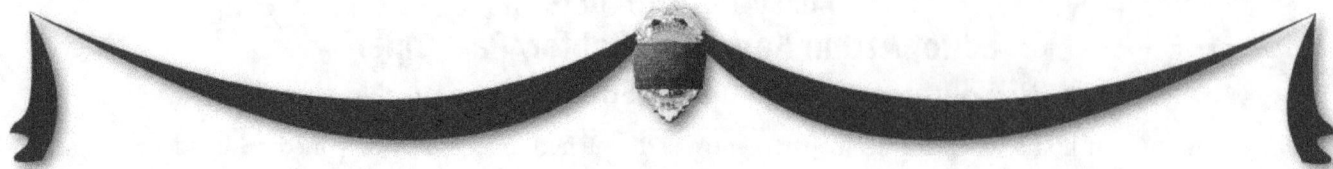

January 26, 2010–1215 hrs
Henry Sandy, Firefighter
Age 52, Volunteer • Northside Volunteer Fire Department, Arkansas

Firefighter Sandy and his spouse, also a firefighter, were at the fire station when the Northside Volunteer Fire Department was dispatched to a residential structure fire. Firefighter Sandy and his spouse responded to the scene in an engine.

A working fire in a vacant residence was found upon arrival. Firefighter Sandy advanced two preconnected hoselines to the front door of the structure and charged the lines for use by arriving firefighters. Firefighter Sandy carried a positive-pressure fan to the residence, started the fan, refueled the fan, and restarted the fan after refueling.

Firefighter Sandy stepped off of the front porch of the residence and suddenly collapsed. Firefighters began CPR and ALS care was provided by the crew of an ambulance that arrived minutes later. Firefighter Sandy was transported to the hospital. He was pronounced dead approximately 45 minutes after his collapse. The cause of death was listed as a heart attack.

For additional information regarding this incident, please refer to NIOSH Fire Fighter Fatality Investigation and Prevention Program Report F2010-26 (www.cdc.gov/niosh/fire/reports/face201026.html).

February 8, 2010–1500 hrs
John B. Coyle, Jr., Assistant Fire Chief
Age 63, Volunteer • West Pend Oreille Fire District, Idaho

Assistant Chief Coyle was helping with the construction of a new fire station for his fire department. He suffered a heart attack while working at the site and later died.

February 10, 2010–1545 hrs
Stanley Lee Giles, Fire Chief
Age 69, Volunteer • Linn Valley Lakes Fire Department, Kansas

The Linn Valley Lakes Fire Department received a mutual-aid request for a tanker (tender) to respond to a structure fire. The tanker responded from the station with one firefighter aboard. Chief Giles responded to the incident but remained at the fire station.

The tanker was cancelled while responding and returned to the fire station. As the tanker backed into the fire station, Chief Giles was crushed between the moving tanker and a parked apparatus. He was pronounced dead at the scene.

For additional information regarding this incident, please refer to NIOSH Fire Fighter Fatality Investigation and Prevention Program Report F2010-07 (www.cdc.gov/niosh/fire/reports/face201007.html).

February 12, 2010–1333 hrs
Donald Gary Mellott, Fire Police Captain
Age 62, Volunteer • Woolrich Volunteer Fire Company No. 1, Pennsylvania

Fire Police Captain Mellott was dispatched to provide scene safety and traffic control services for a mutual-aid motor vehicle crash with entrapment and wires down. He parked his vehicle on the side of the road with its warning lights in operation and set cones and a flare in the roadway to warn approaching drivers.

A civilian vehicle disregarded the road closure, drove over cones, and struck Fire Police Captain Mellott as he faced away from oncoming traffic. He was transported to the hospital but was pronounced dead upon arrival.

For additional information regarding this incident, please refer to NIOSH Fire Fighter Fatality Investigation and Prevention Program Report F2010-06 (www.cdc.gov/niosh/fire/reports/face201006.html).

February 20, 1010–1614 hrs
Jonathan Siemers, Fire Chief
Age 44, Career • Clay Center Fire Department, Kansas

Chief Siemers was the first to arrive at a fire alarm incident in a multiple residential occupancy at 1614 hours. The building was one block from Chief Siemers's residence and Chief Siemers walked to the scene in full turnout gear. Chief Siemers climbed to the third floor of the building and learned that the alarm was accidentally caused by cooking.

Continued on next page.

Arriving firefighters ventilated the building to clear the smoke. Chief Siemers cleared the fire alarm panel, sent firefighters back to quarters, and walked back to his residence. When Chief Siemers arrived home, he complained to his wife of shoulder pain and left a family outing early because he was not feeling well.

At approximately 0414 hours the next morning, firefighters and EMS responders were called to Chief Siemers's home for a medical emergency. Chief Siemers had suffered a heart attack. Firefighters, law enforcement officers, and EMS responders provided ALS care and transported Chief Siemers to the hospital. Chief Siemers was pronounced dead at the hospital at 0450 hours.

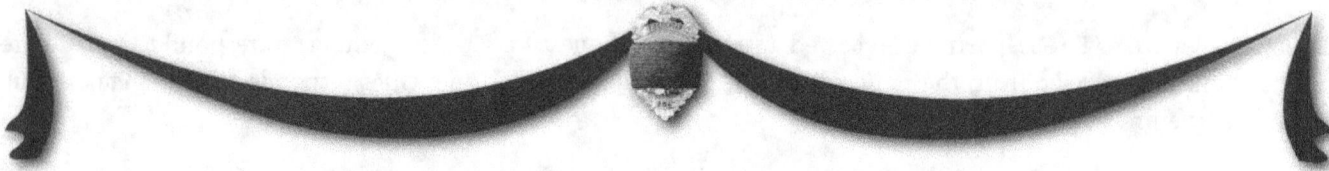

March 1, 2010–2130 hrs
Brian P. Waynant, Sr., Firefighter/Inspector
Age 45, Career • Wilmington Fire Department, Delaware

Firefighter Waynant was on duty. He was struck by a southbound Amtrak train. Delaware State Police, along with Amtrak police and the city of Wilmington, investigated the circumstances surrounding the death. No foul play was suspected.

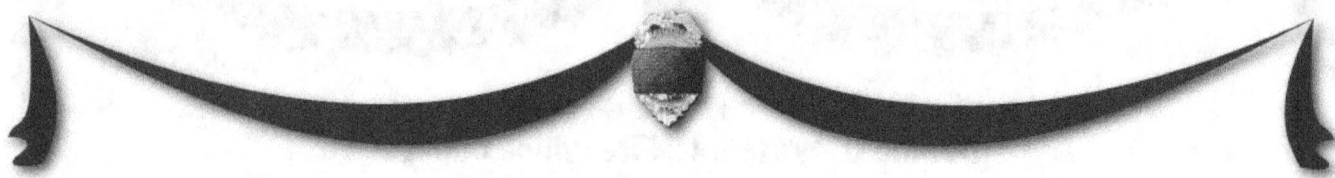

March 5, 2010–2215 hrs
Brian Joseph Rowe, Fire Chief
Age 67, Volunteer • West Forks Volunteer Fire Department, Maine

At 2153 hours, Chief Rowe and the members of his fire department were dispatched to the scene of a snowmobile crash in their jurisdiction. Chief Rowe responded in his personal vehicle, a four-wheel drive, because he knew that it would be needed to access the scene.

Chief Rowe drove approximately 2 miles toward the scene of the incident on an unplowed path. At approximately 2215 hours, he became ill and called other firefighters by radio to request help. Firefighters and EMS responders arrived at his location by snowmobile at 2230 hours. Chief Rowe was in his vehicle and was experiencing difficulty breathing. He was moved to the passenger seat and a firefighter began to back his vehicle down the path. At some point, Chief Rowe lost consciousness and CPR was begun in the moving vehicle.

A medical helicopter dispatched to the earlier crash responded to provide treatment and transportation for Chief Rowe. The helicopter crew assisted with treatment to no avail. Chief Rowe was pronounced dead at 2320 hours.

March 6, 2010–2230 hrs
Gerald Marcheterre, Firefighter
Age 50, Volunteer • Borodino Fire Department, New York

Members of the Borodino Fire Department were responding to a report of a barn that had collapsed under the weight of recent snow. Shortly after leaving the fire station on the response, Firefighter Marcheterre began to cough and had difficulty breathing.

Firefighters requested that an ambulance meet them at a fire station along their route. When the apparatus arrived at the fire station, Firefighter Marcheterre was removed and found to be in respiratory and cardiac arrest. CPR was provided by firefighters and Firefighter Marcheterre was transported to the hospital by ambulance. Firefighter Marcheterre was pronounced dead at the hospital. The cause of death was a heart attack.

March 10, 2010–0508 hrs
Kevin J. Swan, Fire Police Captain
Age 68, Volunteer • Beacon Hose Co. No. 1, Beacon Falls, Connecticut

Fire Police Captain Swan and the members of his fire department were dispatched to a possible structure fire at 0508 hours. Fire Police Captain Swan reported his response by radio.

At 0811 hours, Captain Swan was discovered, deceased, in his personal vehicle in the driveway of his residence. His death was caused by a heart attack.

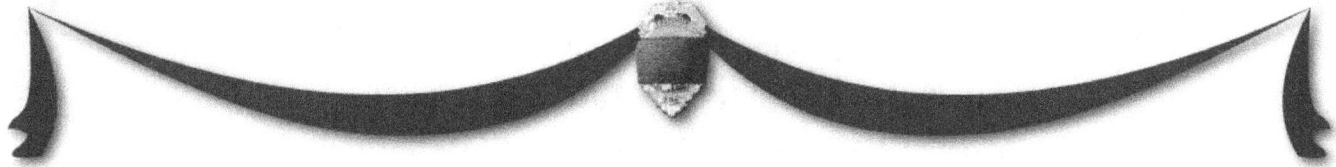

March 13, 2010–0630 hrs
Donald Willard "Donnie" Adkins, III, Firefighter
Age 32, Volunteer • Glasgow Volunteer Fire Department, West Virginia

Firefighter Adkins was a member of a five-person swiftwater rescue team that was deployed to the southern part of West Virginia to assist with swiftwater operations. His deployment was under the direct order of the West Virginia Office of Emergency Services.

Firefighter Adkins, along with two other crew members, was actively engaged in rescue operations when the boat was captured by strong current, struck a bridge, and capsized. All three members of the crew were thrown into the rapidly moving Beaver Creek. Firefighter Adkins did not surface. His body was recovered 6 days later about 6 miles downstream of the incident site. Firefighter Adkins drowned.

For additional information regarding this incident, please refer to NIOSH Fire Fighter Fatality Investigation and Prevention Program Report F2010-09 (www.cdc.gov/niosh/fire/reports/face201009.html).

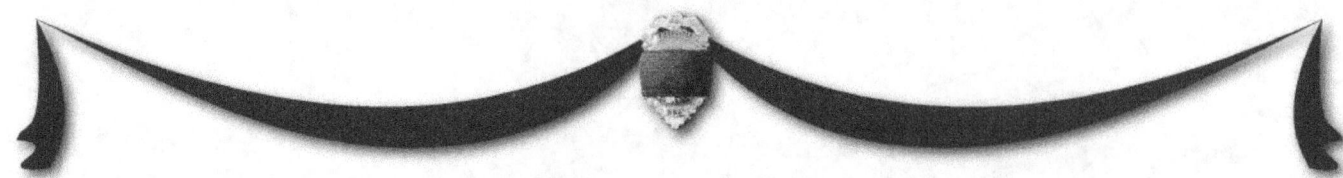

March 21, 2010–1503 hrs
Jeremy Gordon Bolick, Firefighter
Age 23, Volunteer • Blowing Rock Fire Department, North Carolina

Charles Thomas "Tommy" Wright, Firefighter
Age 20, Volunteer • Blowing Rock Fire Department, North Carolina

Firefighter Bolick was the driver and Firefighter Wright was the front seat passenger in a 2008 Ford Mustang. The firefighters were returning from a fire certification class at a local community college.

Firefighter Bolick lost control of the vehicle. It crossed the center line, spun around twice, and was struck by a vehicle in the oncoming lane. Firefighter Bolick and Firefighter Wright were trapped in the vehicle and died of multiple traumatic injuries. Both firefighters were wearing seatbelts and the vehicle's airbags deployed. The law enforcement report on the incident cited excessive speed and poor road conditions as factors in the crash.

The passenger in the other car involved in the crash died 11 days after the crash as a result of injuries received in the incident.

Law enforcement investigators were quoted in the press saying that the crash involved street racing between the vehicle driven by Firefighter Bolick and firefighters in another vehicle. Speeds during the race on the rain-slicked highway topped 100 miles per hour (mph). In December of 2010, an 18-year-old firefighter was charged with one count of spontaneous speed competition and one count of misdemeanor death by vehicle.

March 29, 2010–1610 hrs
John Perry Moore, Firefighter/Paramedic
Age 56, Career • Columbus Division of Fire, Ohio

Firefighter/Paramedic Moore was assigned as a paramedic instructor. He worked 24-hour shifts and was based at the department's training complex. When firefighters noticed that his vehicle had not moved, they entered his quarters to investigate and found that Firefighter/Paramedic Moore had died. His death was caused by a heart attack.

Brian Colin "Boo" Carey, Firefighter
Age 28, Career • Homewood Fire Department, Illinois

Firefighter Carey was on duty when he and his crew were dispatched to a residential structure fire. Law enforcement officers on the scene reported a working fire with a person trapped inside.

When firefighters arrived on scene, they found a large home with an obvious working fire to the rear. Firefighter Carey and two other firefighters advanced a 2-1/2-inch hoseline through the front door of the home. Once inside, they found heavy black smoke about 4 feet off of the floor. As the line was advanced, the smoke level dropped to knee-level.

Firefighters conducting search activities observed fire advancement and yelled to the hoseline crew to evacuate. Once firefighters exited, they found that Firefighter Carey and another firefighter were still in the structure. Fire conditions inside had changed dramatically. The other firefighter was able to make it to within a few feet of the door as he was pulled from the structure by other firefighters.

Firefighters entered the structure with a hoseline to search for Firefighter Carey. He was found wrapped up in the ruptured 2-1/2-inch handline and was not wearing his facepiece, hood, or helmet. Firefighter Carey was removed from the building.

Once outside, Firefighter Carey was treated by other firefighters and transported to the hospital by ambulance. He was pronounced dead at the hospital.

At autopsy, the carboxyhemoglobin level in Firefighter Carey's blood was found to be 30 percent. Firefighter Carey had been wearing his facepiece when he entered the structure. It is unknown why he removed his equipment.

The 87-year-old occupant of the home also perished in the fire. Firefighter Carey was the first LODD in the 109-year history of the Homewood Fire Department.

For additional information regarding this incident, please refer to NIOSH Fire Fighter Fatality Investigation and Prevention Program Report F2010-10 (www.cdc.gov/niosh/fire/reports/face201010.html).

March 30, 2010–1500 hrs
Dennis Wayne Robinson, Captain
Age 61, Career • Three Points Fire District, Arizona

Captain Robinson was assigned as the Training Officer for the Three Points Fire District. On March 30, 2010, training activities included a roof ventilation and overhaul training session.

The training exercise was scheduled to begin at 1300 hours. The preparations for the exercise required Captain Robinson to take sheets of plywood and sections of drywall and transport those pieces to the top of the training trailer.

As the training exercise began, Captain Robinson alternated between providing direction to the firefighters engaged in the exercise, transporting additional wood and drywall onto the training prop, moving fire equipment, and bringing water to the firefighters engaged in the exercise.

As the exercise was conducted, Captain Robinson began to feel ill. At approximately 1400 hours, he went to his office in the fire station to attempt to recover. By 1430 hours, he decided that it would be best for him to go home to recover.

As he drove home in his fire department vehicle, Captain Robinson felt progressively more ill. He stopped at Fire Station 302 and called for assistance on his cell phone. The fire unit and ambulance assigned to that station were not in quarters at the time. In response to his call, firefighters were dispatched to the station to provide assistance.

Responding firefighters found Captain Robinson in extreme pain. He was given oxygen and pain medication by firefighters and transported to the hospital by ambulance.

Captain Robinson's condition continued to deteriorate upon his arrival. As time passed, he was diagnosed as having suffered a massive stroke. With no hope of recovery, life support was removed and Captain Robinson died at approximately 0618 hours on March 31, 2010.

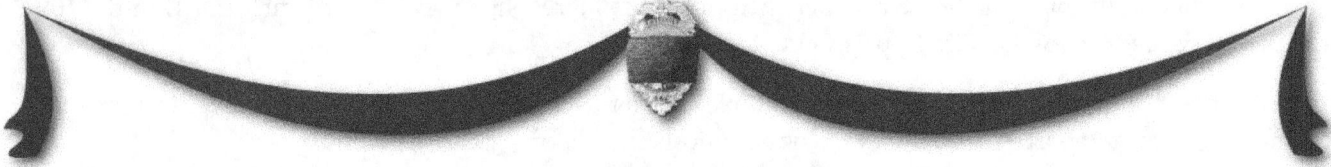

March 31, 2010–1400 hrs
Edward Damian Teare, Lieutenant
Age 53, Career • Independence Fire Department, Ohio

Lieutenant Teare and the other firefighters assigned to his fire station participated in a biannual firefighting skills training session on the morning of March 31, 2010. The exercise included simulated firefighting tasks and was performed in full structural protective clothing, including a self-contained breathing apparatus (SCBA).

Once he completed the evolution, Lieutenant Teare complained of shoulder pain and retired to his quarters. Firefighters were dispatched on an emergency and left the station. Lieutenant Teare did not respond.

When firefighters returned to the station, they found Lieutenant Teare unconscious in his dorm room. He was treated and transported to the hospital where he was pronounced dead due to a heart attack.

For additional information regarding this incident, please refer to NIOSH Fire Fighter Fatality Investigation and Prevention Program Report F2010-11 (www.cdc.gov/niosh/fire/reports/face201011.html).

April 3, 2010–1220 hrs
Leo Andrew Powell, Captain
Age 74, Volunteer • Morgan Township Volunteer Fire Department, Ohio

Captain Powell and members of his fire department responded to a mutual-aid wildland fire at 1220 hours. Captain Powell was the driver/operator of a brush truck. All Morgan Township units were back in service at 1432 hours. At approximately 1745 hours, Captain Powell became ill at his residence and could not be revived. His death was caused by a heart attack.

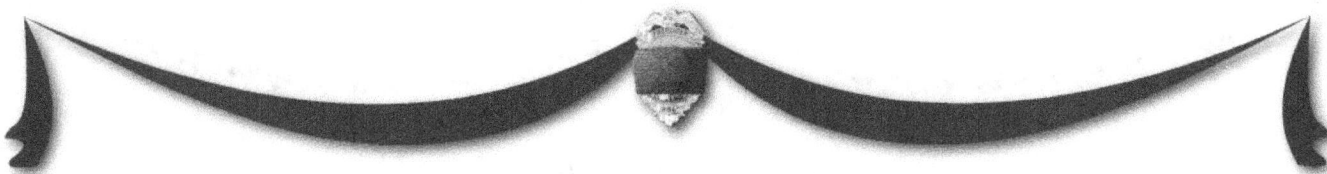

April 9, 2010–1737 hrs
Donald Duaine Schaper, Firefighter
Age 53, Volunteer • Timber Knob Volunteer Fire Department, Missouri

Firefighter Schaper responded to an emergency in his community. As the incident was resolved and firefighters prepared to leave the scene, Firefighter Schaper grabbed his head and chest and collapsed. He was transported to the hospital and was pronounced dead upon his arrival. His death was caused by a heart attack.

April 11, 2010–1555 hrs
Harold D. Reed, Sr., Firefighter
Age 74, Volunteer • Peru Fire District #3, Kansas

Firefighter Reed was assisting with wildland firefighting operations when he became ill. He was transported to the hospital by ambulance but did not survive. His death was caused by a heart attack.

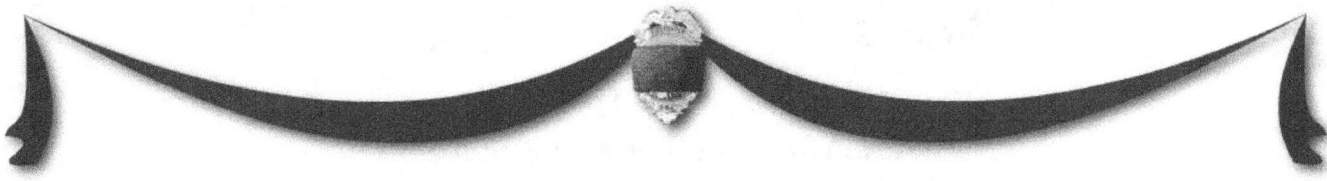

April 11, 2010–0914 hrs
Garrett William Loomis, Assistant Fire Chief
Age 26, Volunteer • Sackets Harbor Fire Department, New York

Assistant Chief Loomis and the members of his fire department responded to a report of a fire in a 20x60 foot oxygen limiting silo containing approximately 20 feet of high moisture corn. The fire may have been caused by embers from an earlier barn fire.

Continued on next page.

Upon arrival, firefighters found a silo with bottom and top doors open, light smoke, and burning embers visible inside the bottom of the silo. They formed a plan to close all silo doors and introduce carbon dioxide to smother the fire.

Assistant Chief Loomis climbed to the top of the silo using a ladder chute attached to the silo. He closed and latched the doors at the top of the silo, reported the completion of his task by radio, and began to climb back down the ladder.

An explosion occurred and Assistant Chief Loomis was thrown 20–30 feet to the ground. Assistant Chief Loomis was transported to the hospital by ambulance. He was pronounced dead at the hospital due to traumatic injuries.

For additional information regarding this incident, please refer to NIOSH Fire Fighter Fatality Investigation and Prevention Program Report F2010-14 (www.cdc.gov/niosh/fire/reports/face201014.html).

April 12, 2010–2000 hrs
Vincent Angelo Iaccino, Captain
Age 65, Volunteer • Roosevelt Fire District Engine Company #1, Hyde Park, New York

Members of the Roosevelt Fire District were attending training at the Dutchess County Fire Training Center. The company drill involved familiarization with newly acquired bailout systems and SCBA training.

Captain Iaccino was involved in SCBA confidence course training wearing full structural firefighting protective clothing, including an SCBA. Captain Iaccino completed the training and was directed to rehab.

While in rehab, Captain Iaccino became ill. He was observed rubbing his chest and complained of chest pains. He was treated by firefighters and by standby EMS responders. Captain Iaccino was transported to the hospital by ambulance. He was pronounced dead shortly after arriving at the emergency room. His death was caused by a heart attack.

For additional information regarding this incident, please refer to NIOSH Fire Fighter Fatality Investigation and Prevention Program Report F2010-22 (www.cdc.gov/niosh/fire/reports/face201022.html).

April 21, 2010–2130 hrs
Steven Scott Crannell, Firefighter
Age 47, Volunteer • Guthrie Center Fire Department, Iowa

Firefighter Crannell participated in a training exercise with members of his department. An old barn was used for live fire training. The training involved hose lays, defensive firefighting operations, and the use of portable water tanks. Training was concluded and Firefighter Crannell went home at approximately 2130 hours.

At 0005 hours, firefighters and EMS responders were dispatched to Firefighter Crannell's home. He had suffered a heart attack. Firefighter Crannell was transported to the hospital by ambulance but died a short time later.

May 1, 2010–1230 hrs
John Polimine, Firefighter

Age 50, Volunteer • Scalp Level & Paint Volunteer Fire Company, Pennsylvania

Firefighter Polimine was participating in an Essentials of Firefighting program. The training included interior and exterior firefighting evolutions. The day started at 0800 hours with student briefings and the collection of baseline vitals on each student.

Firefighter Polimine was the nozzle person for an interior attack evolution. He completed fire extinguishment and ventilation tasks. His crew was instructed to back out of the room. Firefighter Polimine was sluggish and was escorted out of the burn room by an instructor.

EMS staff in rehab evaluated Firefighter Polimine and started an IV to rehydrate him. Once the IV was completed, Firefighter Polimine was reevaluated and found to be feeling better. Firefighter Polimine was allowed to leave rehab after approximately 2 hours to participate in exterior firefighting evolutions.

Firefighter Polimine participated in a propane fire simulation as the nozzle person. At the completion of the exercise, he was directed by an instructor to drink some water as he took a break.

As the next round of exercises was started, Firefighter Polimine could not be located. A search of the area found Firefighter Polimine unconscious in a portable restroom. Firefighters and instructors removed Firefighter Polimine to the ground and began medical treatment.

An AED was applied and delivered three shocks. Firefighter Polimine was transported by ALS ambulance to the hospital where he later died. His death was due to a heart attack.

For additional information regarding this incident, please refer to NIOSH Fire Fighter Fatality Investigation and Prevention Program Report F2010-21 (www.cdc.gov/niosh/fire/reports/face201021.html).

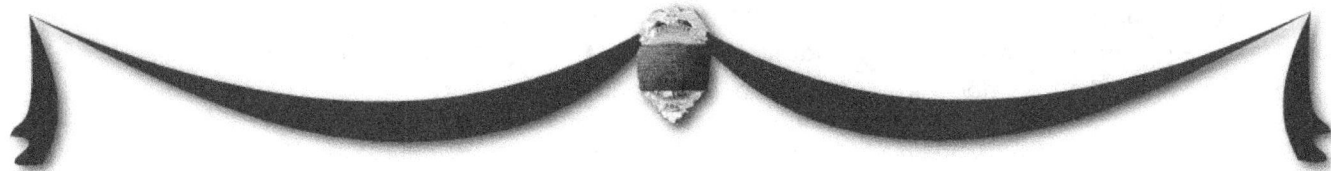

May 19, 2010–0142 hrs
Paul Edward Johnson, Fire Chief

Age 63, Volunteer • Crow Roost Fire Department, Oklahoma

The Crow Roost Fire Department received notification from the local dispatch center of a structure fire involving a house in the Crow Roost Fire District. Chief Johnson responded to the fire station from his residence in his personal vehicle.

Chief Johnson was the first member of the fire department to arrive at the station after receiving the call. Chief Johnson entered the fire station, placed his protective firefighting clothing on Crow Roost Pumper 2, and opened the manual apparatus bay doors.

Continued on next page.

Another firefighter arrived, boarded the apparatus, and the unit responded from quarters. During the response, Chief Johnson reported that he did not feel well. Shortly thereafter, Chief Johnson slowed the truck and told the firefighter to "take the wheel." Chief Johnson then slumped back in the driver's seat. The firefighter was able to reach the brake pedal and brought the truck to a stop in the middle of the road. Chief Johnson was now unresponsive. The firefighter transmitted a Mayday on the truck radio, and then called the local dispatch center, via cell phone, to request a ground ambulance to respond to their location at approximately 0200 hours.

Other firefighters arrived and performed a rapid medical assessment on Chief Johnson. They found that he was not breathing and confirmed that he had no pulse. One firefighter immediately began CPR while another retrieved an AED and oxygen from an apparatus compartment. The AED delivered a shock but did not restore a pulse. As a result, CPR was continued.

An ALS ambulance arrived and transported Chief Johnson to the hospital. Care was continued in the emergency room until Chief Johnson was pronounced dead at 0311 hours. His death was caused by a heart attack.

May 22, 2010–2055 hrs
John Bradford Glaser, Firefighter
Age 33, Career • Shawnee Fire Department, Kansas

The Shawnee Fire Department received a residential fire alarm report at 2052 hours and dispatched a single company. Additional calls reporting a working fire generated the response of a full alarm assignment of three engines, two quints, a Command Officer, and a medic unit.

Firefighters arrived on the scene and found a working fire in a 6,000-square-foot residence. Firefighter Glaser was assigned search and rescue of the structure with his company officer. Bystanders reported that a dog and an elderly couple might be inside of the structure. Firefighter Glaser broke out a side window by the front door and unlocked the door. Heavy black smoke began to pour out of the broken window as he worked.

Firefighter Glaser and his company officer entered the structure and began a search. As they searched, Firefighter Glaser advanced an uncharged 1-3/4-inch handline and a Thermal Imaging Camera (TIC). They located the family dog and carried it to the front door, where it was handed to other firefighters.

Firefighter Glaser and his company officer reentered the structure to continue the search. They were joined by another crew. Shortly thereafter, Firefighter Glaser's company officer was heard calling for him. The second crew scanned the area with their TIC in an attempt to find Firefighter Glaser.

Rapid Intervention Team (RIT) crews were assigned and searched the structure for Firefighter Glaser. After approximately 10 minutes of searching by multiple crews, Firefighter Glaser was located in a small room behind a closed door. He was removed from the structure by firefighters and treated. Firefighter Glaser was transported to the hospital by ambulance where he was pronounced dead.

Investigation revealed that Firefighter Glaser had become ill and vomited into his SCBA facepiece. When he was found, he was lying on his back without his helmet, gloves, and facepiece. His death was caused by smoke inhalation.

For additional information regarding this incident, please refer to NIOSH Fire Fighter Fatality Investigation and Prevention Program Report F2010-13 (www.cdc.gov/niosh/fire/reports/face201013.html).

May 22, 2010–2114 hrs
Kurt Michael Meusel, Firefighter
Age 25, Volunteer • Scales Mound Fire Protection District, Illinois

Firefighter Meusel responded on his 4-wheel drive all-terrain vehicle (ATV) to a search operation for an elderly man with dementia that was reported lost in a wooded area. As he responded, Firefighter Meusel struck a deer in the roadway and was thrown from his vehicle.

A passing motorist reported the crash. Firefighters responded and provided treatment for Firefighter Meusel but he was pronounced dead at the scene. His death was caused by blunt force trauma to the head. Firefighter Meusel was not wearing a helmet at the time of the crash.

The subject of the original search arrived home from a walk as firefighters spoke with his spouse.

May 23, 2010–0210 hrs
David Joseph Irr, Fire Captain
Age 49, Career • Yuma Rural/Metro Fire Department, Arizona

Captain Irr worked overtime from approximately 1945 hours on May 21, 2010, until his regular shift began at 0700 hours on May 22, 2010. During the shift, Captain Irr conducted water supply training, assisted crew members with maintenance of a folding water tank, and responded to two emergency incidents.

Captain Irr was heard sending an incident report by fax at approximately 0210 hours on May 23, 2010. The next morning at shift change, firefighters found Captain Irr deceased in his bunk. The cause of death was a heart attack.

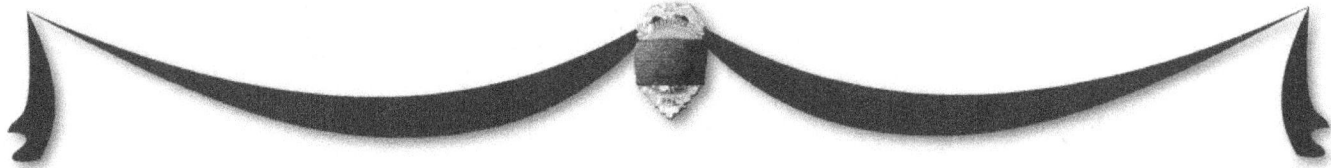

May 29, 2010–Time Unknown
Donald A. Schneider, Jr., Firefighter/Fire Inspector
Age 63, Volunteer • Belleville Fire Department, Wisconsin

Firefighter Schneider was completing paperwork associated with his duties as the department's fire inspector in the fire station. He suffered a heart attack and died. He was discovered some time later by other firefighters.

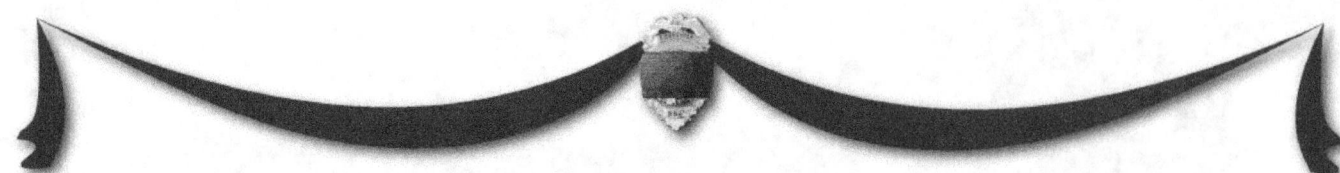

June 1, 2010–1503 hrs
Edward J. Eckert, Firefighter

Age 71, Volunteer • Stafford Township Volunteer Fire Company #1, New Jersey

Firefighter Eckert and the members of his fire department were dispatched to a smoke condition in a residence. Firefighter Eckert rose from a chair to respond and collapsed. His speech was slurred and he did not have a memory of falling.

Firefighters responded to Firefighter Eckert's residence and assisted EMS responders with care. Firefighter Eckert was transported by ambulance to the hospital and then transported by air ambulance to a regional-care facility. Firefighter Eckert died on June 6, 2010, as the result of a CVA.

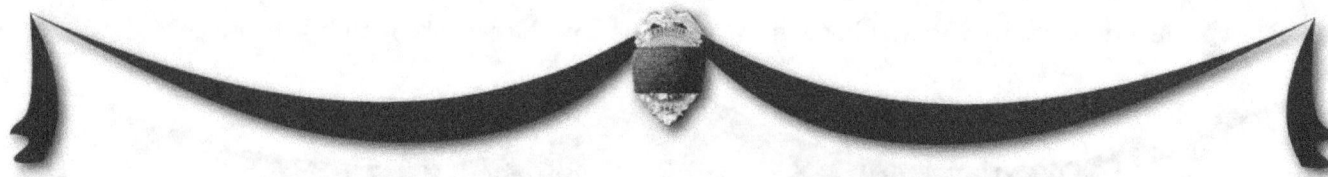

June 19, 2010–Time Unknown
Scott W. Davis, Firefighter

Age 46, Career • Oswego Fire Department, New York

Firefighter Davis and other firefighters responded to an emergency medical incident that concluded at 2300 hours on June 18, 2010. He went off duty the next morning and traveled with his family. Firefighter Davis suffered a heart attack later in the day on June 20, 2010, and died.

June 22, 2010–1900 hrs
Terrell Gene Nielsen, Sr., Firefighter

Age 56, Volunteer • Bryan County Fire Department, Georgia

Firefighter Nielsen attended training at the fire station on June 22, 2010. During the course of the training exercise, Firefighter Nielsen struck his elbow on a portion of the body of an engine as he removed SCBA cylinders for use in the training. He noted his injury to other firefighters but thought that it was insignificant. Firefighter Nielsen signed out of training at 2100 hours.

At approximately 0300 hours the next morning, Firefighter Nielsen awoke and began vomiting. He was experiencing pain is his injured elbow. An ambulance was called. Arriving EMS responders treated the incident as a cardiac event based on Firefighter Nielsen's symptoms and transported him to the hospital.

Continued on next page.

Firefighter Nielsen was diagnosed as suffering from necrotizing fasciitis (commonly known as flesh eating disease). He underwent multiple surgeries to attempt to control the spread of the disease and was transferred to a burn unit for treatment on June 26, 2010.

Firefighter Nielsen underwent a total of 18 operations during 79 days in the hospital. Although his condition seemed to improve for a time, he died on September 9, 2010, as a result of complications related to the injury.

June 23, 2010–1530 hrs
Chet D. Bauermeister, Fire Chief
Age 46, Volunteer • Franklin County Fire District 4, Washington

Chief Bauermeister and the members of his fire department responded to a mutual-aid wildland fire. Chief Bauermeister and another firefighter staffed a converted snowcat-type ATV equipped with a water tank and pump.

As the snowcat ascended a hill, it lost traction and began to tumble down the slope. The passenger was ejected during the roll. Chief Bauermeister remained in the vehicle as it rolled 3.5 times. He was pinned in the wreckage. Firefighters and emergency medical technicians (EMTs) gained access to the wreckage and assessed Chief Bauermeister's condition and found him to be deceased. He died of multiple traumatic injuries.

Neither occupant of the snowcat was wearing a seatbelt at the time of the incident.

For additional information regarding this incident, please refer to NIOSH Fire Fighter Fatality Investigation and Prevention Program Report F2010-15 (www.cdc.gov/niosh/fire/reports/face201015.html).

June 27, 2010–1800 hrs
Cecil Jay Brown, Fire Chief
Age 42, Volunteer • Gresston Volunteer Fire Department, Georgia

After the passage of a thunderstorm, Chief Brown and the members of his fire department were dispatched to a roadway in their district that was blocked by a large fallen tree. Chief Brown responded to the scene and assisted firefighters in removal of the tree, including the use of a chain saw.

Chief Brown left the scene suddenly in his personal vehicle. He encountered another firefighter and motioned for the firefighter to follow him. Chief Brown pulled off of the road and got into the passenger seat of his vehicle. Chief Brown was having difficulty breathing and asked the firefighter to drive him to the hospital.

When the firefighter noticed that Chief Brown had become unconscious, he pulled off the road, removed Chief Brown from his vehicle, and started CPR. EMS responders arrived, provided treatment, and transported Chief Brown to the hospital. Chief Brown died as the result of a heart attack.

June 30, 2010–1140 hrs
Charles Phillip Hornberger, Engineer
Age 60, Volunteer • Milmont Park Fire Company Station #49, Pennsylvania

Engineer Hornberger responded to his fire station when his department was dispatched to an automatic fire alarm. When he arrived at the station, he attempted to get into the driver's seat of the engine for the response but felt dizzy and told other firefighters that he was going to standby at the station.

At approximately 1200 hours, the incident was concluded and Engineer Hornberger went home. At 1515 hours, Engineer Hornberger suffered a heart attack at home. He was transported to the hospital but did not recover. He died as a result of the heart attack on July 12, 2010.

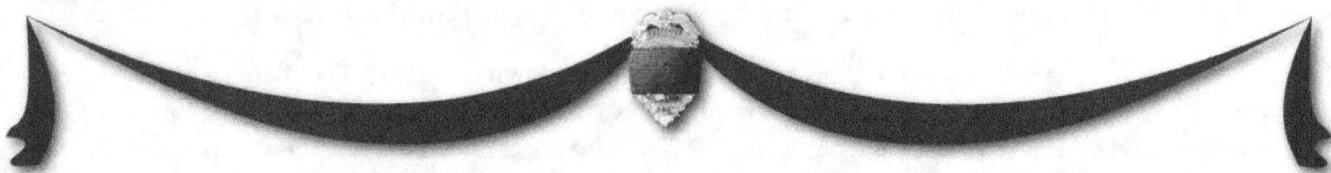

July 1, 2010–1045 hrs
Charles Robert "Bob" Flintom, Engineer
Age 61, Career • Pelham Batesville Fire Department, South Carolina

Engineer Flintom had been on light duty for some time due to medical issues. Engineer Flintom's duties were limited to in-station duties including apparatus checks, SCBA maintenance, and station cleaning duties.

At approximately 1045 hours, Engineer Flintom experienced a fall in the fire station. Firefighters treated him for a cut on his head and recommended that he go to the hospital to be checked. Engineer Flintom agreed to call his wife and have her take him to the emergency room.

Firefighters checked on Engineer Flintom in the hospital later that night and found him in better condition than before. Engineer Flintom, however, later took a turn for the worse and died on July 4, 2010. His death was caused by a head injury received in the fall at the fire station.

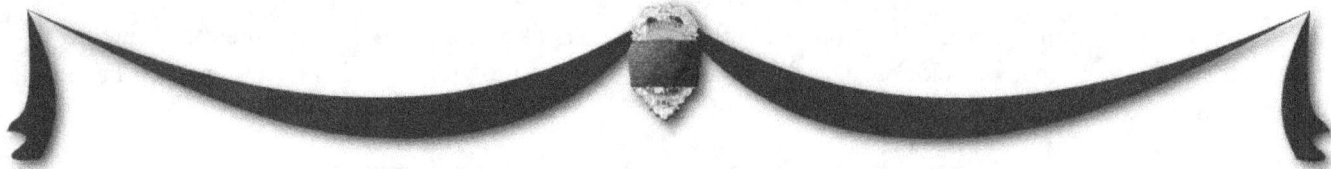

July 1, 2010–1310 hrs
Frank William Fouts, V, Lieutenant
Age 37, Career • City of Kankakee Fire Department, Illinois

Lieutenant Fouts responded to several emergency medical and fire incidents while working his shift from June 30, 2010, to the morning of July 1, 2010. At 1310 hours on July 1, 2010, Lieutenant Fouts was discovered unconscious in the back yard of his home. He was unable to be revived and was pronounced dead of a heart attack.

For additional information regarding this incident, please refer to NIOSH Fire Fighter Fatality Investigation and Prevention Program Report F2010-23 (www.cdc.gov/niosh/fire/reports/face201023.html).

July 3, 2010–2141 hrs
Thomas T. Araguz, III, Captain
Age 30, Volunteer • Wharton Volunteer Fire Department, Texas

Captain Araguz and his engine company arrived approximately 20 minutes after dispatch and found a working fire in a large egg production and processing facility. Captain Araguz and another captain advanced an attack line into the structure. The captains became separated from the hoseline and then from each other in the interior.

The other captain with Captain Araguz was able to find a metal exterior wall and banged on it until firefighters on the exterior were able to cut open the wall to allow his escape. The fire progressed and attempts to locate Captain Araguz were unsuccessful. He died of thermal injuries and smoke inhalation.

The Texas State Fire Marshal's Office prepared a detailed report on this incident. The report is available at www. tdi.state.tx.us/reports/fire/documents/fmloddaraguz.pdf

July 9, 2010–1030 hrs
Douglas Lee Smith, Firefighter/Chief Driver
Age 50, Volunteer • Liberty Hose Company No. 1, Williamstown, Pennsylvania

Firefighter Smith arrived at his fire station to respond to a structure fire. He took his position in the driver's seat of the apparatus and began to exhibit signs of illness. The officer of the apparatus called for emergency medical assistance.

Firefighter Smith was treated at the fire station and went into cardiac arrest while undergoing treatment. He was transported to the hospital but was pronounced dead upon arrival. His death was caused by a heart attack.

July 14, 2010–1545 hrs
Richard Lawrence Springman, Firefighter
Age 20, Volunteer • Trout Run Volunteer Fire Company, Pennsylvania

Firefighter Springman was responding to a fatal crash on a local highway. As he approached the scene, he lost control of his pickup truck and struck a guardrail. News reports stated that Firefighter Springman oversteered, causing his truck to travel sideways into the guardrail. The vehicle rolled, ejecting Firefighter Springman. The vehicle came to rest approximately 300 yards from the original crash incident scene. Firefighter Springman was not wearing a seatbelt and was killed due to multiple traumas.

July 23, 2010–1830 hrs
Steven Nelson Costello, Lieutenant
Age 46, Career • Burlington Fire Department, Vermont

Lieutenant Costello was on duty and responded to three emergency incidents during the first part of his shift. At 1800 hours, his company returned to quarters from their last response. Lieutenant Costello donned his workout clothing and told his crew that he was going to work out.

Approximately 15 minutes later, Lieutenant Costello was discovered beside the treadmill, unconscious. Firefighters provided CPR and applied an AED. Lieutenant Costello was transported to the hospital by ambulance. He died on July 30, 2010, as the result of a heart attack.

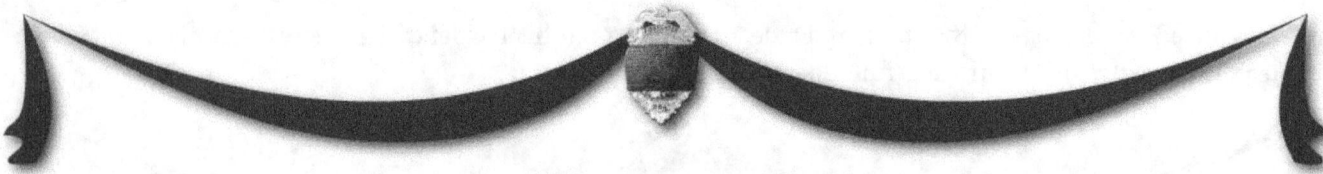

July 25, 2010–1537 hrs
David A. Sullivan, Firefighter
Age 70, Volunteer • Otis Fire Department, Massachusetts

Firefighter Sullivan and the members of his fire department assisted with a structure fire in a neighboring community on July 24, 2010. The last Otis unit cleared at 1815 hours. Firefighter Sullivan died of a heart attack the next day.

July 24, 2010–1553 hrs
Steven John Velasquez, Lieutenant
Age 40, Career • Bridgeport Fire Department, Connecticut

Michel Baik, Firefighter
Age 49, Career • Bridgeport Fire Department, Connecticut

Lieutenant Velasquez and Firefighter Baik were on duty at their regular assignment at Ladder 11. The unit was dispatched at 1549 hours to respond to a residential structure fire.

Upon their arrival at the scene, firefighters found an active fire on the second floor of the residence. Lieutenant Velasquez and Firefighter Baik were assigned to go to the third floor of the building to look for fire extension and to search for victims. Lieutenant Velasquez and Firefighter Baik proceeded to the third floor as assigned. Both firefighters were wearing full structural protective clothing, including SCBAs.

Continued on next page.

When they arrived on the third floor, Lieutenant Velasquez reported stable conditions by radio to the Incident Commander (IC) and requested a hoseline be brought to the floor. A few moments later, the IC observed smoke coming from the third floor and attempted to contact Lieutenant Velasquez by radio to no avail.

The IC ordered an evacuation of the building by all firefighters in order to account for everyone on the scene. In the process of the evacuation, firefighters were observed on the third floor in distress.

The RIT was deployed to the third floor for rescue. Lieutenant Velasquez and Firefighter Baik were found unconscious and brought to the exterior of the building.

Firefighters and emergency responders immediately began to provide medical treatment for both firefighters. Lieutenant Velasquez was transported by ambulance to Saint Vincent's Hospital where he was pronounced dead. Firefighter Baik was transported by ambulance to Bridgeport Hospital where he was also pronounced dead.

Autopsies and toxicological studies were performed on both firefighters. Lieutenant Velasquez and Firefighter Baik both died of smoke inhalation suffered as they fought the structure fire.

For additional information regarding this incident, please refer to NIOSH Fire Fighter Fatality Investigation and Prevention Program Report F2010-18 (www.cdc.gov/niosh/fire/reports/face201018.html).

July 26, 2010–1630 hrs
Posey Wayne Dillon, Fire Chief
Age 59, Volunteer • Rocky Mount Fire Department #1, Virginia

William Daniel "Danny" Altice, Firefighter
Age 67, Volunteer • Rocky Mount Fire Department #1, Virginia

Chief Dillon and Firefighter Altice were responding to a mutual-aid structure fire incident. Their engine entered an intersection on a red light and was struck by a Sport Utility Vehicle (SUV). The collision caused Chief Dillon to lose control of the apparatus and it rolled.

Both firefighters were ejected from the vehicle. Neither firefighter was wearing a seatbelt. Both firefighters died due to traumatic injuries.

For additional information regarding this incident, please refer to NIOSH Fire Fighter Fatality Investigation and Prevention Program Report F2010-19 (www.cdc.gov/niosh/fire/reports/face201019.html).

August 2, 2010–1025 hrs
Christopher Wayne Adams, Forest Ranger I
Age 25, Wildland Full-Time • Arkansas Forestry Commission

Forest Ranger Adams worked a fire scene on the morning of August 2, 2010. His activities consisted of plowing lines around a previous fire to assure containment. Once this was completed, Forest Ranger Adams was assigned to drive a tractor trailer truck hauling a D5 Caterpillar plow.

As Forest Ranger Adams approached a curve in the road, he lost control of the vehicle; it left the roadway and overturned. Forest Ranger Adams was trapped in the cab of the truck and was pronounced dead at the scene. Forest Ranger Adams was wearing his seatbelt at the time of the crash.

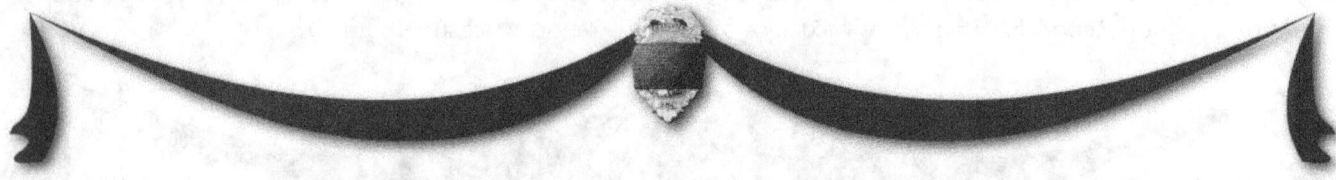

August 9, 2010–0030 hrs
Christopher D. Wheatley, Firefighter/Paramedic
Age 31, Career • Chicago Fire Department, Illinois

Firefighter Wheatley and his ladder company were dispatched to a structural fire in a four-story commercial and residential structure. Upon their arrival, firefighters observed smoke and sparks coming from a cooking exhaust fan chute.

Firefighter Wheatley ascended a ladder attached to the side of the building to gain access to the roof. He was wearing full structural firefighting protective clothing, including a SCBA. He also carried a water fire extinguisher with him as he climbed.

When Firefighter Wheatley reached the roof, he lost his grip and fell 53 feet to the ground. Firefighter Wheatley landed on his feet and immediately dropped to the ground. Firefighter Wheatley was treated by firefighters and transported to the hospital. He was pronounced dead at the hospital due to multiple injuries. The most significant injury noted in the autopsy was transaction of the aorta.

For additional information regarding this incident, please refer to NIOSH Fire Fighter Fatality Investigation and Prevention Program Report F2010-25 (www.cdc.gov/niosh/fire/reports/face201025.html).

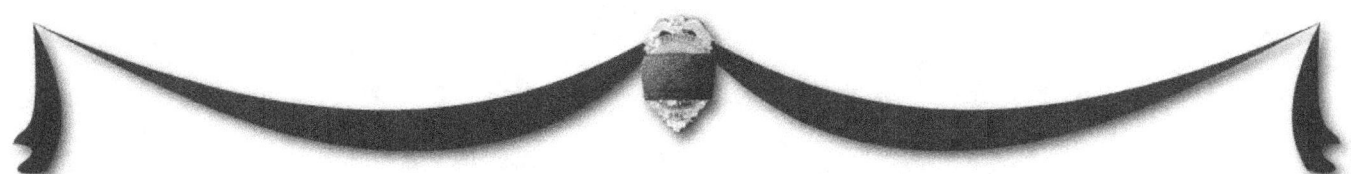

August 19, 2010–2045 hrs
Jonathan Lewis "Johnny" Littleton, Captain
Age 40, Volunteer • Pine Level Volunteer Fire Department, North Carolina

Captain Littleton and other firefighters participated in a technical rescue class on the evening of August 19, 2010. The class began at the Pine Level Fire Department at 1830 hours and was concluded at 2230 hours. The class involved rope work and was very physically demanding.

Captain Littleton participated in all facets of the class but began to feel ill toward the end of class. Instructors allowed him to sit out the class cleanup duties. Captain Littleton was scheduled to work that evening in his job with the town police department but had gotten someone else to cover his shift.

Captain Littleton went home and stayed in bed for the better part of the next day. His wife last saw him at approximately 1500 hours. At approximately 2030 hours, Captain Littleton's daughter found him deceased in bed. Firefighter and EMS responders provided treatment but Captain Littleton had died of a heart attack.

August 22, 2010–2121 hrs
Douglas Farrington, Firefighter
Age 44, Volunteer • Delta Cardiff Volunteer Fire Company, Pennsylvania

On August 22, 2010, Firefighter Farrington responded to his fire station for a mutual-aid barn fire. He missed the response but remained in the fire station to staff a backup engine. Firefighter Farrington was at the fire station for about an hour and assisted other firefighters in cleaning and putting the equipment back in service when they returned from the response. Firefighter Farrington went home at approximately 2230 hours.

The next day, Firefighter Farrington worked in the morning at his full-time job as a fire inspector and then left work to complete a mowing job. At 1105 hours, Firefighter Farrington momentarily lost consciousness. Another firefighter assisting with the mowing job called for EMS. When the ambulance arrived, Firefighter Farrington had gone into respiratory and cardiac arrest. He was pronounced dead at 1219 hours after arriving at the hospital. His death was caused by a heart attack.

September 3, 2010–Time Unknown
Larry W. Suiter, Fire Chief
Age 65, Volunteer • Lorraine Green Garden Fire Department, Kansas

Chief Suiter responded to an anhydrous ammonia leak at a local grain elevator on the evening of September 3, 2010. He died at his home in the early hours of September 4, 2010, of an apparent heart attack.

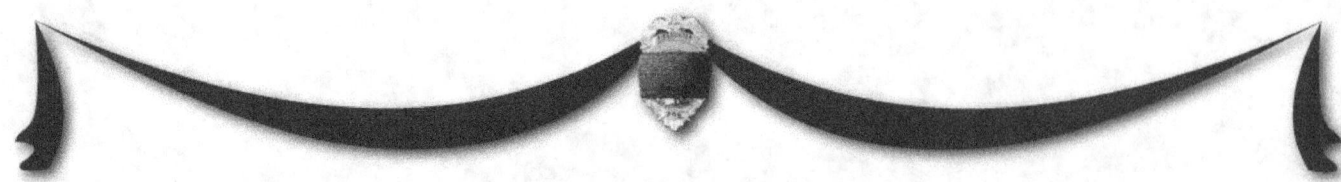

September 6, 2010–1837 hrs
John Kelly, Firefighter/Chief Driver
Age 51, Volunteer • Tarrytown Fire Department, New York

A resident of the Village of Tarrytown complained that the sewer service to their home was clogged. The Department of Public Works (DPW) attempted to clear the clog in the area of the home but was not successful. The Tarrytown Fire Station was in the area of the sewer problems. There are three sewer system openings in proximity to the fire station: one in front of the station, one in the station, and one behind the station. DPW called the fire chief and asked him to open the fire station so that DPW employees could have access to the sewer system inside of the station.

Firefighters assisted DPW employees with access to the opening behind the fire station. A DPW employee entered the sewer opening. The DPW employee lost consciousness as he descended the interior ladder and fell to the bottom. The DPW foreperson requested assistance from firefighters. The fire chief asked dispatch for EMS assistance and also started a fire department incident.

As an atmospheric meter from the fire department was prepared for use, Firefighter Kelly entered the opening. He was not wearing a SCBA or harness. When Firefighter Kelly was half way down the ladder, he lost consciousness and fell to the bottom. Additional resources were called to the scene and Firefighter Kelly and the DPW employee were removed. Both Firefighter Kelly and the DPW employee died of asphyxiation due to an oxygen deficient atmosphere.

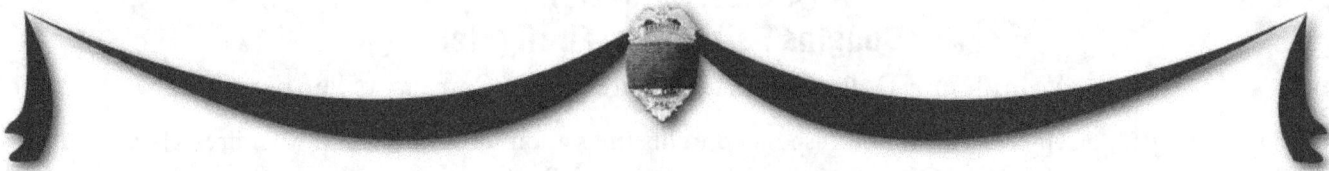

September 7, 2010–2340 hrs
Donald Ray Lam, Jr., Forest Ranger Technician III
Age 58, Wildland Full-Time • Kentucky Division of Forestry

Forest Ranger Lam was working with other firefighters to contain the Scotts Chapel Road Fire in Livingston County. Forest Ranger Lam was clearing a fire break for containment at the base of a bluff when a burning snag broke loose on top and rolled downhill over a small bluff, striking Lam from behind.

Firefighter Lam was knocked out and sustained a serious head injury, fractured hip, bruises, and second-degree burns on his calves from the snag. The impact left him unconscious and with serious injuries, including the burns from which he did not recover. Forest Ranger Lam died on February 17, 2011.

September 14, 2010–0930 hrs
Edward Avoughn Mosley, Firefighter
Age 66, Volunteer • Steele Creek Acres Volunteer Fire Department, Morgan, Texas

In March of 2010, Steele Creek Volunteer Fire Station One was destroyed by fire. Members of the fire department were working on the construction of a new fire station. Firefighter Mosley was injured on the worksite. He fell backwards and struck his head on concrete. Firefighter Mosley died as the result of his injury on September 26, 2010.

September 16, 2010–1127 hrs
James Michael Owen, Firefighter/Paramedic
Age 56, Career • Orange County Fire Authority, California

Firefighter Owen and other firefighters were participating in a day of technical rescue training at the Orange County Fire Authority's Regional Fire Operations and Training Center. The day began with classroom instruction and stretching exercises prior to the start of training evolutions.

The first training evolution began at 1055 hours. Teams of firefighters, including Firefighter Owen, moved 4,000-pound cubes of concrete using only hand tools. Firefighter Owen and his group successfully completed the evolution. At 1120 hours, firefighters were released for lunch.

As Firefighter Owen walked toward his apparatus, he suddenly collapsed. Firefighters immediately provided CPR and ALS-level care. By 1131 hours, Firefighter Owen was in an ambulance en route to the hospital. Firefighter Owen was pronounced dead at 1222 hours in the hospital emergency room, despite all efforts in the field and in the hospital to revive him. His death was caused by a heart attack.

For additional information regarding this incident, please refer to NIOSH Fire Fighter Fatality Investigation and Prevention Program Report F2010-34 (www.cdc.gov/niosh/fire/reports/face201034.html).

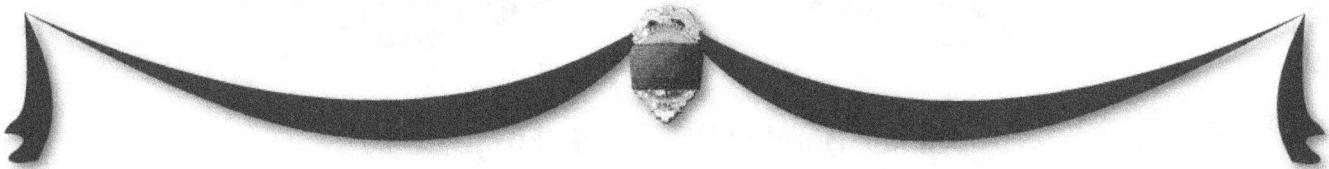

September 20, 2010–1650 hrs
Mark Phillip Johnson, Deputy Chief
Age 55, Volunteer • Hinsdale Fire Department, Illinois

Deputy Chief Johnson responded to a medical emergency incident on September 20, 2010. The incident concluded at 1148 hours and Deputy Chief Johnson returned to his office. At approximately 1545 hours, Deputy Chief Johnson told other firefighters that he was going to the basement of the fire station to work out.

Continued on next page.

At 1655 hours, firefighters discovered Deputy Chief Johnson collapsed behind an exercise machine. Firefighters began treatment and transported Deputy Chief Johnson to a local hospital. He could not be revived and was pronounced dead at 1739 hours. His death was caused by a heart attack.

For additional information regarding this incident, please refer to NIOSH Fire Fighter Fatality Investigation and Prevention Program Report F2010-33 (www.cdc.gov/niosh/fire/reports/face201033.html).

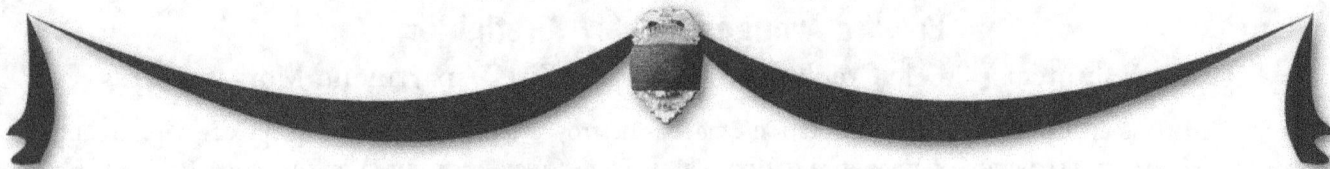

September 24, 2010–1800 hrs
Ryan Neil Seitz, Firefighter
Age 26, Volunteer • McArthur Fire Department, Ohio

Firefighter Seitz and other firefighters from Ohio responded to a group of wildland fires in Eastern Ross County, Ohio. Firefighter Seitz was staffing a brush truck. The water tank on the brush truck failed under pressure and Firefighter Seitz was struck in the chest by a piece of the tank. He was killed as a result of trauma and was pronounced dead at the scene.

September 24, 2010–1436 hrs
Ronald W. Stephan, Firefighter
Age 61, Volunteer • Lynn Volunteer Fire Department, Indiana

On September 24, 2010, Firefighter Stephan responded to a mutual-aid wildland fire. By 1620 hours, his department was back in service. The next morning, Firefighter Stephan suffered a heart attack while working on his farm. EMS was called at 0950 hours. Firefighter and EMS responders provided CPR and ALS-level care. Their efforts were not successful and Firefighter Stephan was pronounced dead due to a heart attack.

September 24, 2010–1756 hrs
William Harold "Hal" Clark, Firefighter
Age 54, Volunteer • Atlantic Volunteer Fire & Rescue Company, Virginia

Firefighter Clark and the members of his department were assisting with wildland fires in New Church, VA. Firefighter Clark operated a 300 foot 1-1/2-inch hoseline on the fire for approximately 35 minutes. Firefighter Clark became ill while working the incident. He was treated at the scene and taken to the hospital. He was pronounced dead at the hospital as the result of a heart attack.

For additional information regarding this incident, please refer to NIOSH Fire Fighter Fatality Investigation and Prevention Program Report F2010-35 (www.cdc.gov/niosh/fire/reports/face201035.html).

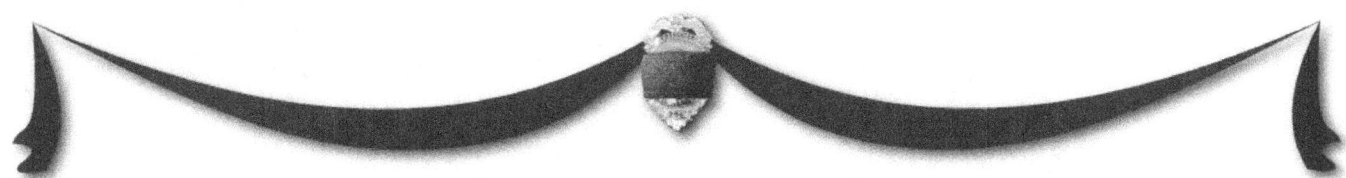

September 27, 2010–1846 hrs
Robert Hall, Firefighter

Age 57, Volunteer • Lynchburg Area Joint Fire & Ambulance District, Ohio

Firefighter Hall attended a meeting at the fire station. As he left the fire station, he became ill. Firefighters and EMS responders provided assistance and Firefighter Hall was transported to the hospital. Firefighter Hall did not recover and later died as the result of a CVA.

October 2, 2010–2245 hrs
Thomas Dale "Tinker Tom" Innes, Assistant Fire Chief

Age 61, Volunteer • Hindsboro Community Fire Protection District, Illinois

Assistant Chief Innes responded to an EMS call at 2245 hours. On the scene, he assisted with patient treatment and then helped restore equipment to readiness for the next incident. He returned home at approximately 2330 hours. Assistant Chief Innes did not feel well and found that his blood pressure was high.

Assistant Chief Innes and his wife decided that it would be best for him to go to the emergency room by personal vehicle. As Mrs. Innes drove, Assistant Chief Innes became unresponsive. An ambulance was called and transported Assistant Chief Innes to the hospital. He did not recover and was later pronounced dead due to a heart attack.

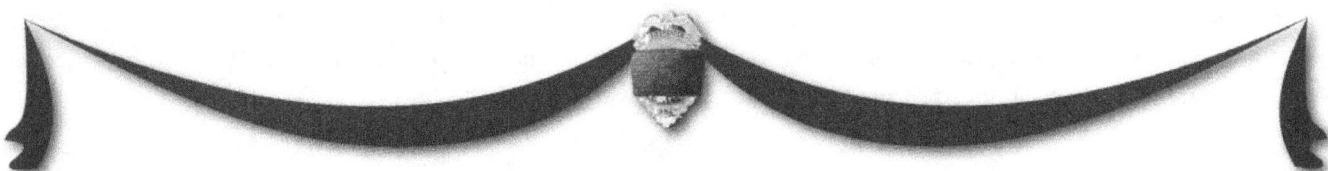

October 3, 2010–2006 hrs
James Carlyle Saunders, Firefighter

Age 52, Career • Sacramento Metropolitan Fire District, California

Firefighter Saunders and other firefighters from his fire department controlled a wildland fire that occurred near a local golf course. Firefighters were in the process of mopping up when Firefighter Saunders became ill. He was treated by firefighters and transported by ambulance to the hospital. Firefighter Saunders suffered a heart attack and died on October 7, 2010.

October 19, 2010–1515 hrs
Daniel C. Wilson, Firefighter
Age 58, Volunteer • Jerusalem Township Fire Department, Ohio

On October 19, 2010, at approximately 1515 hours, Firefighter Wilson responded to an EMS incident. At the conclusion of the response, he returned home. It was noted that he appeared to be feeling poorly when he left. On October 20, 2010, at approximately 1000 hours, he requested EMS to his residence due to feeling ill. He was transported to the hospital where he underwent emergency heart surgery and remained hospitalized until his death on October 23, 2010.

October 19, 2010–2115 hrs
William E. "Billy" Akin, Jr., Fire Police Captain
Age 54, Volunteer • Ghent Volunteer Fire Company Number One, New York

Fire Police Captain Akin and the members of his fire department were dispatched to an automobile crash with injuries on a local roadway. Fire Police Captain Akin was responding is his personal vehicle when he suffered a heart attack.

Fire Police Captain Akin's vehicle accelerated through an intersection and crashed into a utility pole with sufficient force to break the pole and drop live electrical wires to the street. Firefighters responding to the original call came upon the crash, removed Fire Police Captain Akin from his vehicle, and provided treatment. He was transported to the hospital but was pronounced dead shortly after his arrival.

October 24, 2010–1800 hrs
Randall Scott Davenport, Firefighter
Age 37, Career • Marshall Fire Department, Missouri

Firefighter Davenport responded to a residential structure fire involving a 1,200-square-foot single-story home. The structure was well-involved and Firefighter Davenport assisted with extinguishment and overhaul once the fire was knocked down. The incident was dispatched at 1307 hours and he returned to quarters at approximately 1400 hours.

At 1539 hours, Firefighter Davenport was dispatched to a mutual-aid structure fire. Firefighter Davenport drove a tanker (tender) and worked onscene to protect an exposure that was threatened. He returned to the fire station at approximately 1800 hours.

Continued on next page.

Firefighter Davenport complained to other firefighters that he was fatigued but did not complain of any cardiac symptoms. Firefighters watched a sports game on television until midnight. Firefighter Davenport slept in a chair in the station's day room.

At approximately 0600 hours, firefighters found Firefighter Davenport deceased in the chair. He had suffered a heart attack during the night and died.

October 26, 2010–1530 hrs
Bruce M. Bachinsky, Lieutenant
Age 47, Career • Waterbury Fire Department, Connecticut

Lieutenant Bachinsky went off duty at 0800 hours on October 26, 2010, after responding to multiple incidents. As he was riding his bicycle at approximately 1630 hours, he suffered a heart attack.

A passerby found Lieutenant Bachinsky at the roadside. A bystander performed CPR until law enforcement and EMS responders arrived. An AED was used and Lieutenant Bachinsky was taken to the hospital where he was later pronounced dead.

October 30, 2010–0615 hrs
Kevin Daniel Quinn, Lieutenant
Age 52, Career • Dayton Fire Department, Ohio

Lieutenant Quinn was the onduty representative for the Dayton Fire Department in the Montgomery County Regional Dispatch Center. During his 24-hour shift, Lieutenant Quinn monitored emergency activity and participated in physical fitness training activities. He also completed his annual SCBA fit test.

At approximately 2230 hours on October 29, 2010, Lieutenant Quinn retired to his quarters for a break. A dispatcher that entered his quarters at approximately 0620 hours on October 30, 2010, found him deceased. His death was caused by a heart attack.

October 31, 2010–2045 hrs
Gary Lowell Cummins, Firefighter

Age 61, Volunteer • Brocton Fire Protection District, Illinois

Firefighter Cummins and the members of his fire department were dispatched to a structure fire in their district. Firefighter Cummins drove an engine apparatus to the scene. When he arrived on the scene, he was assigned by the fire chief to position his apparatus to support firefighting operations.

As Firefighter Cummins backed the apparatus into position, he suffered a heart attack. The apparatus, still in reverse, left the roadway, backed through a corn field, jumped a ditch, and came to rest in another ditch. Firefighters secured the apparatus and removed Firefighter Cummins from the vehicle.

CPR was initiated and Firefighter Cummins was transported to the hospital by ambulance. He was pronounced dead upon arrival.

November 1, 2010–1445 hrs
Richard E. "Rick" Drake, II, Lieutenant

Age 39, Volunteer • German Township Fire Department, Indiana

Lieutenant Drake responded to an EMS incident that involved difficulty breathing. He provided patient care and assisted with lifting the patient for transport by ambulance. The incident concluded at 1224 hours.

After their return to the fire station, Lieutenant Drake began to perform the weekly check on apparatus. His wife was with him at the station. At approximately 1445 hours, Lieutenant Drake began to experience chest pains and other signs of an acute heart attack. His wife called 9-1-1 for assistance and Lieutenant Drake also requested help by radio.

Firefighter and EMS responders arrived at the station and provided assistance. Lieutenant Drake was transported to the hospital by ambulance but did not survive.

November 5, 2010–1830 hrs
Leonard Arthur Murray, Jr., Fire Captain
Age 53, Volunteer • Jackson Township Fire Department, Indiana

A firefighter/mechanic was making steering system repairs to a brush truck inside of the Jackson Township Fire Department maintenance building. The mechanic turned the key to unlock the steering column. The brush truck accidentally lunged forward.

At the same time the brush truck lunged forward, Captain Murray and another firefighter were walking in front of the vehicle. The other firefighter was knocked clear but Captain Murray was crushed between the brush truck and a wall. He received a head injury and was killed instantly.

For additional information regarding this incident, please refer to NIOSH Fire Fighter Fatality Investigation and Prevention Program Report F2010-37 (www.cdc.gov/niosh/fire/reports/face201037.html).

November 10, 2010–0245 hrs
James C. Gumbert, Firefighter
Age 63, Volunteer • North Irwin Volunteer Fire Department, Pennsylvania

Firefighter Gumbert responded to his fire station for a mutual-aid structure fire. When he arrived at the fire station, he was experiencing chest pains and collapsed. Firefighters provided CPR in the fire station and he was transported by ambulance to the hospital. He was pronounced dead shortly after his arrival.

November 13, 2010–1415 hrs
Chance Hyatt Zobel, Firefighter
Age 23, Career • Columbia Fire Department, South Carolina

Firefighter Zobel and his engine company were dispatched to a fire in the median of a local highway. When his company and another engine arrived on the scene, they positioned their apparatus in a shielding position to protect firefighters from traffic.

As Firefighter Zobel and other firefighters were extinguishing the fire, a two-vehicle crash occurred on the highway behind the apparatus. The crash caused one of the vehicles to slide between the guardrail and one of the engines. The vehicle entered the fire scene and struck Firefighter Zobel and another firefighter. Both firefighters were transported to the hospital. Firefighter Zobel died due to traumatic injuries.

November 19, 2010–1421 hrs
Worne T. Hall, Captain
Age 86, Volunteer • Hitchens Volunteer Fire Department, Kentucky

Captain Hall and the members of his fire department responded to a vehicle crash involving an ATV. Captain Hall's department requested mutual aid from another fire department that brought an ATV capable of patient transport. Once the patient was assessed, EMS responders requested a medical helicopter for transport.

Captain Hall was directed to set up a helispot for the medical helicopter at a local school. As the helicopter landed, Captain Hall collapsed of an apparent heart attack. He was treated by firefighters and EMS responders, including members of the helicopter crew, and transported to the hospital by ambulance. He could not be revived and was pronounced dead at the hospital.

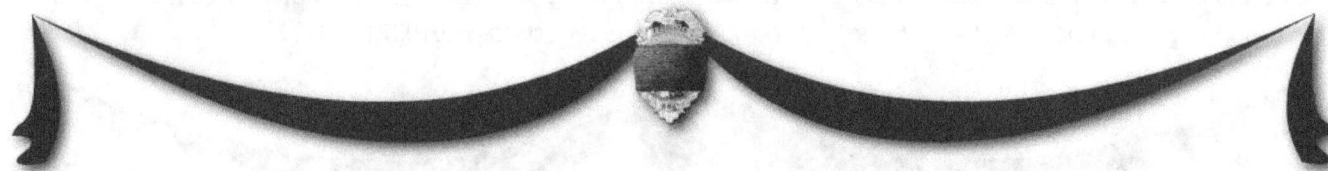

November 23, 2010–1420 hrs
Fernando Julio Sanchez, Inmate Firefighter
Age 25, Wildland Part-Time • California Department of Corrections

Firefighter Sanchez was a passenger in a wildland crew carrier. There were as many as 12 firefighters in the vehicle. An SUV crossed the centerline of the highway and crashed into the crew carrier. Several firefighters were ejected in the crash and a number received severe injuries. Firefighter Sanchez was killed in the crash.

November 25, 2010–2130 hrs
Kenneth Marshall, Jr., Firefighter
Age 33, Paid-on-Call • Rehoboth Fire Department, Massachusetts

Firefighter Marshall and the members of his fire department were dispatched to a kitchen fire. Firefighter Marshall responded to the fire station and was the driver of Engine 2 for the incident. As the apparatus responded, firefighters noticed that Firefighter Marshall slumped over. Firefighters were able to stop the apparatus and remove Firefighter Marshall.

CPR was started and Firefighter Marshall was transported to the hospital, where he was later pronounced dead. The kitchen fire was found to be food on the stove.

November 26, 2010–0021 hrs
Gary M. Valentino, Firefighter
Age 40, Career • Fire Department City of New York, New York

Firefighter Valentino was working in a light-duty assignment as a division driver. He had been on duty for approximately 12 hours when he became tired and went to rest in a bunkroom. Fellow firefighters found him unconscious approximately 3 hours later. Firefighter started CPR and EMS was requested. Firefighter Valentino could not be revived.

December 6, 2010–2000 hrs
Dillon C. Denton, Lieutenant
Age 64, Volunteer • Charlotte Road/Van Wyck Fire Department, South Carolina

Lieutenant Denton participated in a training exercise with other firefighters. Near the end of the exercise, Lieutenant Denton was assisting other firefighters loading hose on an engine. Lieutenant Denton told other firefighters that he did not feel well and he was told to take a rest.

As Lieutenant Denton walked to an area where he could sit down, he stumbled. Firefighters came to his aid and found that he was seriously ill. An ambulance was called and Lieutenant Denton was flown by medical helicopter to a regional medical facility.

Lieutenant Denton died as the result of a CVA at 0130 hours on December 7, 2010.

December 13, 2010–1440 hrs
Jimmy W. Tuberville, Fire Chief
Age 64, Volunteer • Milledgeville Fire Department, Tennessee

A wildland fire was reported in the field across from the Milledgeville fire station. Firefighters responded and used flappers to extinguish the flames. Chief Tuberville drove the department's brush truck to the scene and helped to fight the fire.

Chief Tuberville returned the tools he had used to the truck and was talking with firefighters when he suddenly collapsed. Firefighters found that he had no pulse and began CPR while an ambulance was called. CPR was continued in the ambulance on the way to the hospital. Chief Tuberville was pronounced dead at the hospital.

December 16, 2010–1130 hrs
Chad Lee Null, Firefighter/Paramedic
Age 33, Career • Sullivan Fire Department, Indiana

At 0955 hours, Firefighter Null and his partner responded to their second EMS incident of the shift. The incident involved a woman that had fallen in a narrow hallway. Firefighter Null and his partner had to maneuver in close quarters to place the patient on a backboard. In addition, the driveway of the home was covered with snow and ice, so Firefighter Null and his partner had to carry the gurney and patient over the lawn to the ambulance. Firefighter Null and his partner arrived back at the fire station at 1057 hours.

Firefighter Null went into the fire station and began to complete reports on the day's incidents. While other firefighters were out of the room for a few moments, Firefighter Null collapsed. Firefighters then entering the room found Firefighter Null and began to provide treatment, including CPR. He was transported to the hospital by ambulance but was not revived. His death was caused by a heart attack.

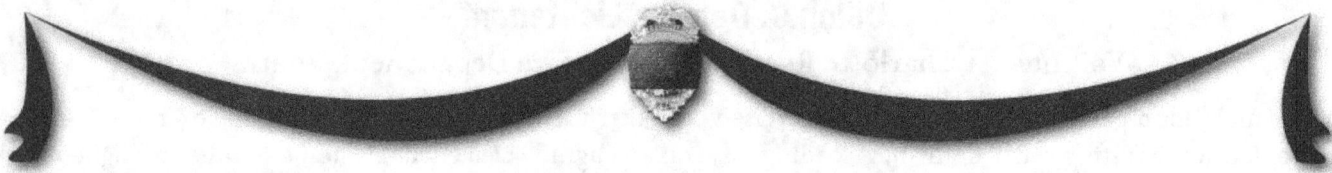

December 22, 2010–0654 hrs
Edward J. Stringer, Firefighter
Age 47, Career • Chicago Fire Department, Illinois

Corey D. Ankum, Firefighter
Age 34, Career • Chicago Fire Department, Illinois

While working a fire at a vacant commercial building, 16 firefighters reportedly became trapped and were injured after the roof of the building collapsed. Two firefighters, Firefighter Edward Stringer and Firefighter Corey Ankum, passed away from injuries sustained in the incident.

Ten other firefighters were reported to be in stable condition while four others were reported to be in critical condition.

For additional information regarding this incident, please refer to NIOSH Fire Fighter Fatality Investigation and Prevention Program Report F2010-38 (www.cdc.gov/niosh/fire/reports/face201038.html).

December 24, 2010–1900 hrs
Thomas G. Hardy, Fire Chief
Age 68, Volunteer • Athens Township Volunteer Fire Department, Michigan

Chief Hardy had just returned to the fire station from a fire call when he suffered a heart attack, fell to the floor, and struck his head. He was treated by firefighters and transported to the hospital. While in the hospital, Chief Hardy's condition worsened and he remained in a coma for several days before passing away from his injuries on December 31, 2010.

December 28, 2010–0930 hrs
Kenneth Adamo, Captain
Age 48, Volunteer • Elmwood Park Fire Department, New Jersey

Firefighter Adamo was on an emergency standby at his fire station for the snow emergency declared by New Jersey Acting Governor Stephen M. Sweeney. During his period of duty, Firefighter Adamo also reportedly responded to one call, a smoke detector sounding.

Upon release from duty, Firefighter Adamo returned home and went to bed. He was found deceased in his bed the following morning from a suspected heart attack.

FIREFIGHTER FATALITIES FROM PREVIOUS YEARS

January 30, 2007–1053 hrs
Donnie Caldwell, Lieutenant
Age 71, Volunteer • Ghent Area Volunteer Fire Department, West Virginia

The propane gas service for a local convenience store was being transferred from one provider to another. Service technicians from the new provider were at the store to install the new propane tank, transfer propane from the old tank to the new tank, and establish service for the store.

At 1042 hours, firefighters were dispatched to the report of a gas leak at this local convenience store. Firefighter Dorsey was on duty and responded as a part of an ambulance crew. Firefighters arrived on the scene to find an uncontrollable propane leak. Firefighters reported seeing a billowing cloud of vapor or mist that was striking the eaves of the building and traveling along the ground. An evacuation was begun.

Continued on next page.

An explosion occurred at approximately 1053 hours. Two firefighters and two gas company employees were killed in the blast. Five other people received serious injuries. Among the injured was Lieutenant Caldwell. He died of complications of the injuries he received in the blast. Lieutenant Caldwell died on May 13, 2010, at the age of 74.

An investigation was completed by the United States Chemical Safety Board. Additional information about this investigation can be found at www.csb.gov/assets/document/CSBFinalReportLittleGeneral.pdf

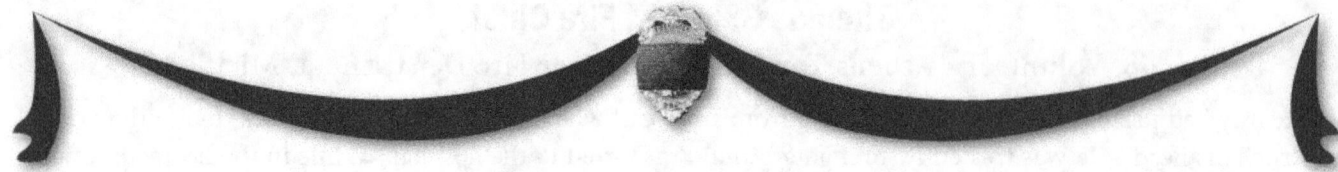

December 27, 2009–1839 hrs
Peter James Coe, Firefighter
Age 43, Volunteer • Shoreham Volunteer Fire Department, Vermont

Firefighter Coe and his family were driving in his personal vehicle when they came upon a car that had run off of the road into the ditch. Firefighter Coe stopped his car and went out to see if the occupants of the vehicle needed assistance.

Firefighter Coe repositioned his vehicle and attached a tow strap between the vehicles to pull the disabled vehicle from the ditch. As he returned to his car, he was struck and killed by a passing vehicle.

APPENDIX B

Firefighter Fatality Inclusion Criteria – National Fire Service Organizations

The National Fire Protection Association (NFPA), the National Fallen Firefighters Foundation (NFFF), the U.S. Fire Administration (USFA), and other organizations individually collect information on firefighter fatalities in the United States. Each organization uses a slightly different set of inclusion criteria that are based, at least in part, on the purposes of the information collection for each organization and data consistency.

As a result of these differing inclusion criteria, statistics about firefighter fatalities may be provided by each organization that does not coincide with one another. This section explains the inclusion criteria for each organization and provides information about these differences.

The USFA includes firefighters in this report that died while on duty, became ill while on duty and later died, or died within 24-hours of an emergency response or training, regardless of whether the firefighter complained of illness while on duty. The USFA counts firefighter deaths that occur in the 50 States, the District of Columbia, and United States protectorates such as Puerto Rico and Guam. Detailed inclusion criteria for this report appear near the beginning of report.

For 2010, the USFA reported 87 onduty firefighter fatalities.

Firefighter Fatalities Resulting from Incidents Occurring in 2010

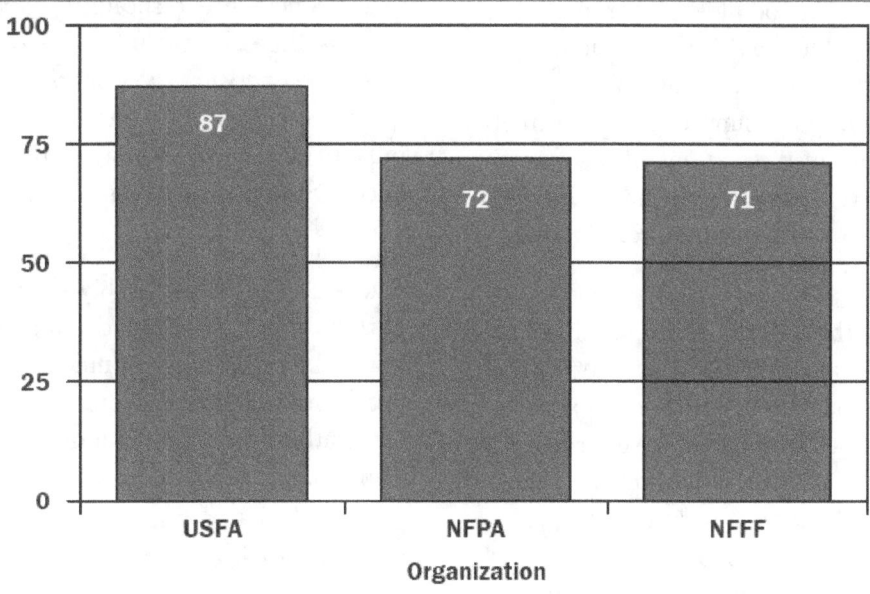

Inclusion Criteria for NFPA's Annual Firefighter Fatality Study

Introduction

Each year, the NFPA collects data on all firefighter fatalities in the United States that resulted from injuries or illnesses that occurred while the victims were onduty. The purpose of the study is to analyze trends in the types of illnesses and injuries resulting in deaths that occur while firefighters are on the job. This annual census of firefighter fatalities in its current format dates back to 1977. (Between 1974 and 1976, NFPA published a study of onduty firefighter fatalities that was not as comprehensive.)

What is a Firefighter?

For the purpose of the NFPA study, the term *firefighter* covers all uniformed members of organized fire departments, whether career, volunteer or combination, or contract; full-time public service officers acting as firefighters; State and Federal government fire service personnel; temporary fire suppression personnel operating under official auspices of one of the above; and privately employed firefighters, including trained members of industrial or institutional fire brigades, whether full or part time.

Under this definition, the study includes, besides uniformed members of local career and volunteer fire departments, those seasonal and full-time employees of State and Federal agencies who have fire suppression responsibilities as part of their job description, prison inmates serving on firefighting crews, military personnel performing assigned fire suppression activities, civilian firefighters working at military installations, and members of industrial fire brigades. Impressed civilians would also be included if called on by the officer in charge of the incident to carry out specific duties. The NFPA study includes fatalities that occur in the 50 States and the District of Columbia.

What does 'on duty' mean?

The term *on duty* refers to being at the scene of an alarm, whether a fire or nonfire incident; being en route while responding to or returning from an alarm; performing other assigned duties such as training, maintenance, public education, inspection, investigations, court testimony, and fundraising; and being oncall, under orders or on standby duty other than at home or at the individual's place of business. Fatalities that occur at a firefighter's home may be counted if the actions of the firefighter at the time of injury involved firefighting or rescue.

Onduty fatalities include any injury sustained in the line of duty that proves fatal, any illness that was incurred as a result of actions while onduty that proves fatal, and fatal mishaps involving nonemergency occupational hazards that occur while onduty. The types of injuries included in the first category are mainly those that occur at an incident scene, in training, or in accidents while responding to or returning from alarms. Illnesses (including heart attacks) are included when the exposure or onset of symptoms are tied to a specific incident of onduty activity. Those symptoms must have been in evidence while the victim was on duty for the fatality to be included in the study.

Fatal injuries and illnesses are included, even in cases where death is considerably delayed. When the onset of the condition and the death occur in different years, the incident is counted in the year of the condition's onset. Medical documentation specifically tying the death to the specific injury is required for inclusion of these cases in the study.

Categories not included in the study

The NFPA study does not include, members of fire department auxiliaries; nonuniformed employees of fire departments; emergency medical technicians (EMTs) who are not also firefighters; chaplains; or civilian dispatchers. The study also does not include suicides as onduty fatalities even when the suicide occurs on fire department property.

The NFPA recognizes that a comprehensive study of firefighter onduty fatalities would include chronic illnesses (such as cardiovascular disease and certain cancers) that prove fatal and that arose from occupational factors. In practice, there is, as of yet, no mechanism for identifying onduty fatalities that are due to illnesses that develop over long periods of time. This creates an incomplete picture when comparing occupational illnesses to other factors as causes of firefighter deaths. This is recognized as a gap the size of which cannot be identified at this time because of the limitations in tracking the exposure of firefighters to toxic environments and substances and the potential long-term effects of such exposures.

2010 Experience

In 2010, a total of 72 onduty firefighter deaths occurred in the United States, according to the NFPA inclusion criteria.

National Fallen Firefighters Foundation

In 1997, fire service leaders formulated new criteria to determine eligibility for inclusion on the National Fallen Firefighter Memorial. Line-of-duty deaths (LODDs) shall be determined by the following standards:

1. (a) Deaths of firefighters meeting the Department of Justice's (DOJ's) Public Safety Officers' Benefits (PSOB) program guidelines, and those cases that appear to meet these guidelines whether or not PSOB staff has adjudicated the specific case prior to the annual National Fallen Firefighters Memorial Service; and

    (b) Deaths of firefighters from injuries, heart attacks, or illnesses documented to show a direct link to a specific emergency incident or department-mandated training activity.

2. While PSOB guidelines only cover public safety officers, the Foundation's criteria also include contract firefighters and firefighters employed by a private company, such as those in an industrial brigade, provided that the deaths meet the standards listed above.

3. Some specific cases will be excluded from consideration, such as deaths attributable to suicide, alcohol or substance abuse, or other gross abuses as specified in the PSOB guidelines.

The National Fallen Firefighters Memorial was built in 1981 in Emmitsburg, MD. The names listed there begin with those firefighters who died in the line of duty that year. The U.S. Congress created the NFFF to lead a nationwide effort to remember America's fallen firefighters. Since 1992, the tax-exempt, nonprofit Foundation has developed and expanded programs to honor our fallen fire heroes and assist their families and coworkers by providing them with resources to rebuild their lives. Since 1997, the Foundation has managed the National Memorial Service held each October to honor the firefighters who died in the line of duty the previous year.

At the October 2011 Memorial Weekend, the Foundation will be honoring 89 firefighters who died in the line of duty. Of those 89 being honored, 71 died in 2010 as the result of incidents that occurred in 2010, and 18 others who died in previous years as the result of incidents that occurred in previous years.

The following section is a listing of the firefighters that will be honored by the Foundation in October of 2011.

Firefighter deaths that occurred in 2010 as the result of an incident that occurred in 2010:

Arizona

David J. Irr
Yuma Rural/Metro Fire Department

Dennis W. Robinson
Three Points Fire District

Arkansas

Christopher W. Adams
Arkansas Forestry Commission

David A. Curlin
Pine Bluff Fire and Emergency Services

Henry Sandy
Northside Volunteer Fire Department

California

James M. Owen
Orange County Fire Authority

Continued on next page.

Fernando J. Sanchez
California Department of Corrections

James C. Saunders
Sacramento Metropolitan Fire District

Connecticut

Bruce M. Bachinsky
Waterbury Fire Department

Michel Baik
Bridgeport Fire Department

Kevin J. Swan
Beacon Hose Co. No. 1

Steven J. Velasquez
Bridgeport Fire Department

Georgia

Cecil J. Brown
Gresston Volunteer Fire Department

Illinois

Corey D. Ankum
Chicago Fire Department

Brian C. Carey
Homewood Fire Department

Gary L. Cummins
Brocton Fire Protection District

Frank W. Fouts, V
City of Kankakee Fire Department

Thomas D. Innes
Hindsboro Community Fire Protection District

Mark P. Johnson
Hinsdale Fire Department

Edward J. Stringer
Chicago Fire Department

Christopher D. Wheatley
Chicago Fire Department

Indiana

Richard E. Drake, II
German Township Fire Department

Chad L. Null
Sullivan Fire Department

Ronald W. Stephan
Lynn Volunteer Fire Department

Iowa

Steven S. Crannell
Guthrie Center Fire Department

Kansas

Stanley L. Giles
Linn Valley Lakes Fire Department

John B. Glaser
Shawnee Fire Department

Harold D. Reed, Sr.
Peru Fire District #3

Jonathan L. Siemers
Clay Center Fire Department

Larry W. Suiter
Lorraine Green Garden Fire Department

Kentucky

Terry L. Cannon
Buechel Fire Protection District

Worne T. Hall
Hitchens Volunteer Fire Department

Maine

Brian J. Rowe
West Forks Volunteer Fire Department

Massachusetts

Kenneth Marshall, Jr.
Rehoboth Fire Department

David A. Sullivan
Otis Fire Department

Michigan

Thomas G. Hardy
Athens Township Volunteer Fire Department

Mississippi

Jerry W. Thompson
Linwood Volunteer Fire Department

Missouri

Randall S. Davenport
Marshall Fire Department

Continued on next page.

Donald D. Schaper
Timber Knob Volunteer Fire Department

New Jersey

Kenneth Adamo
Elmwood Park Fire Department

Edward J. Eckert
Stafford Township Volunteer Fire Company #1

New York

William E. Akin
Ghent Volunteer Fire Company Number One

Scott W. Davis
Oswego Fire Department

Vincent A. Iaccino
Roosevelt Fire District Engine Company #1

John Kelly
Tarrytown Fire Department

LeRoy A. Kemp
Tioga Center Fire Department

Garrett W. Loomis
Sackets Harbor Fire Department

Gerard Marcheterre
Borodino Fire Department

Ohio

Joseph McCafferty
Lancaster Fire Department

Leo A. Powell
Morgan Township Volunteer Fire Department

Ryan N. Seitz
McArthur Fire Department

Edward D. Teare
Independence Fire Department

Daniel C. Wilson
Jerusalem Township Fire Department

Oklahoma

Paul E. Johnson
Crow Roost Fire Department

Pennsylvania

Douglas Farrington
Delta Cardiff Volunteer Fire Company

James C. Gumbert
North Irwin Volunteer Fire Department

Charles P. Hornberger
Milmont Park Fire Company Station #49

Donald G. Mellott
Woolrich Volunteer Fire Company No. 1

John Polimine
Scalp Level & Paint Volunteer Fire Company

Douglas L. Smith
Liberty Hose Company No. 1

Richard L. Springman
Trout Run Volunteer Fire Company

South Carolina

Dillon C. Denton
Charlotte Road/Van Wyck Fire Department

Chance H. Zobel
Columbia Fire Department

Tennessee

Jimmy W. Tuberville
Milledgeville Fire Department

Texas

Thomas T. Araguz, III
Wharton Volunteer Fire Department

Vermont

Steven N. Costello
Burlington Fire Department

Virginia

William D. Altice
Rocky Mount Fire Department #1

William H. Clark
Atlantic Volunteer Fire & Rescue Company

Posey W. Dillon
Rocky Mount Fire Department #1

Washington

Chet D. Bauermeister
Franklin County Fire District 4

West Virginia

Donald W. Adkins, III
Glasgow Volunteer Fire Department

DEATHS FROM PREVIOUS YEARS

Alabama

Steven F. Bouchard
Snowdoun Volunteer Fire Department

John E. Lee, III
City of Pelham Fire Department

California

Bryan K. Zollner
CAL Fire

Florida

Victor B. Scott
Otter Creek Volunteer Fire Department

Michael A. Trullinger
Chattahootchie Volunteer Fire Department

Kansas

Urban A. Eck
Wichita Fire Department

Louisiana

Tommy L. Adams
Shreveport Fire Department

Maryland

Donald W. Hubbel
Baltimore City Fire Department

New York

Erich Lachmann
Washington Heights Fire Department

Salvatore Scarentino
Fire Department of New York

Josef L. Welenofsky
Holtsville Fire Department

North Carolina

Richard A. Miller
Belmont Fire Department

Ohio

Gregory A. Northup
Gallipolis Fire Department

Sammy R. Smith
Village of Antwerp

Oregon

Steven A. Uptegrove
USDA Forest Service, Wallowa-Whitman
National Forest

Pennsylvania

Richard A. Burns
Pittsburgh Bureau of Fire

Tennessee

Timothy A. Byrd
Dover-Stewart County Rescue Squad

Vermont

Peter J. Coe
Shoreham Volunteer Fire Department

www.ingramcontent.com/pod-product-compliance
Lightning Source LLC
Chambersburg PA
CBHW081220170526
45165CB00009B/2887